Contents

PART I The Framework of Negotiations

1. Introduction 1

2. The Givens: geological, technological, economic,
              environmental 4

3. The Notion of the Common Heritage:
              Implications and Critique 31

4. The Participants 43

5. The Negotiating Process 54

PART II The Issues

6. The Geographical Boundary of the Common Heritage 98

7. The System of Exploitation 134

8. Production Control 180

9. The Structure of the Authority 194

10. Dispute Settlement 215

11. Pioneer Investment and the Interim Regime 224

CONCLUSION

12. Outcome and Prospects 236

Bibliography 248

Index Index-1

Acknowledgements

This book attempts to tell the story of one aspect – arguably the most important aspect – of the third United Nations Conference on the Law of the Sea (UNCLOS III). I have been able to watch UNCLOS III in action, and talk to many of its participants, in my capacity as an observer on behalf of the Friends' World Committee for Consultation and the World Federation of United Nations Associations. It was in that way that I attended virtually the whole of the second session, in Caracas in 1974, and the sixth session, in New York in 1977; and parts of the third session in Geneva in 1975, the fourth and fifth sessions in New York in 1976, the Geneva portion of each of the seventh and eighth sessions (1978 and 1979), the resumed eighth session in New York in 1979, the resumed ninth session in Geneva in 1980, the New York portion of the tenth session in 1981, and the adoption session, also in New York, in 1982.

All these visits to UNCLOS III were possible only because of the financial support I have enjoyed from a variety of sources, none, I am glad to say, at all dependent on any of the political or economic interests being promoted at the conference. Some benefactors covered my costs in whole or in part for several visits: The David Davies Memorial Institute for International Studies, the Friends' World Committee for Consultation, the Trustees of the Norman Angell Benefaction to the University of Sussex, (Mr & Mrs E.A. Lane) and the University of Sussex itself. That I was able to attend the sixth session for two weeks in 1976 was entirely due to a grant from the Nuffield Foundation. What was then the Friends Peace and International Relations Committee, (of London Yearly Meeting), the Gerald Bailey Award Fund, the Gilbert Murray Trust Fund, and the United Nations Association of Great Britain and Northern Ireland all contributed to my attending at least one session. I should like to thank all of the foregoing most warmly for their assistance and also, again, the Nuffield Foundation for a further grant to cover the cost of producing a photo-ready manuscript. I also owe further debts to the University of Sussex for enabling me to attend the Law of the Sea Institute's conference in the Hague in 1978 and, much more substantially, the paid leave granted me in Autumn 1973, Summer 1977 and Spring 1979. All these leaves were mainly devoted, in one way or another, to study of, and writing about, the theme of this book.

I should like to thank George Allen and Unwin (Publishers) Ltd., for permission to reproduce extracts from Grotius: Prolegomena to the three books on the Law of War and Peace, taken from M. G. Forsyth, H. M. A. Keens-Soper and P. Savigear (eds), the Theory of International Relations, on p.48. In spite of considerable efforts, I have been unable to discover whether the highly informative and illuminating paper from which Table 2.2 on p.9 is taken, ("A Preliminary Analysis of the World Distribution of Subsea Metal-rich

Internationalizing the Seabed

RODERICK OGLEY
The School of Social Sciences,
The University of Sussex

Gower

Published by
Gower Publishing Company Limited,
Gower House, Croft Road, Aldershot, Hampshire GU11 3HR, England

and

Gower Publishing Company,
Old Post Road, Brookfield, Vermont 05036, U.S.A.

Reprinted 1985

British Library Cataloguing in Publication Data

Ogley, Roderick
  Internationalizing the seabed.
  1. Ocean bottom (Maritime law)   2. Marine
  resources conservation——Law and legislation
  I. Title
  341.45   JX4426

Library of Congress Cataloging in Publication Data

Ogley, Roderick
  Internationalizing the seabed.

  Bibliography:
  Includes index
  1. Ocean bottom (Maritime law). I. Title.
JX4426.035 1984   341.4'5   83-20555

ISBN 0 566 00629 4

Printed and bound in Great Britain by
Biddles Ltd, Guildford and King's Lynn

INTERNATIONALIZING THE SEABED

Dedicated to Miriam Levering,
for her unique blend of vision and humanity

Manganese Nodules," by Vincent McKelvey and the late Nancy Wright) was subsequently published. I have therefore assumed that there was no copyright in it. If I am wrong, I apologise to whoever may own the copyright for not having secured permission to reproduce it.

There are many individuals who by their patience, cooperation and encouragement, have helped to sustain me over this decade of seabed negotiations: my fellow observers, above all Miriam Levering who embodied the Quaker ideal of "answering that of God in every one"; participants in UNCLOS III, both delegates and international civil servants; and many others who have taught, studied, or written about this unique exercise in global law-making. I thank them all, and also those at the University of Sussex who have helped in the process of producing this manuscript; particularly David Hitchin and his team at the Research Data Analysis Unit (Diana Hitchin, Sharon Gretton and Andrew Clews), who guided me, via the University computer, to my first appreciation of the possibilities of word processing, and my secretary Helen Warner, who with Pat Bennett and other secretarial staff of the School of Social Sciences, typed earlier versions of several chapters. Finally, I should like, however inadequately, to record my appreciation of the loving support, and understanding I have had from Anne, my wife, and Simon, Julian and Kate, my children, while I have been absorbed in this protracted enterprise.

Lewes, July 1983

PART I
The Framework of Negotiations

1 Introduction

In December 1982, at Montego Bay, Jamaica, the nations of the world met to sign the Final Act of the largest, longest and conceivably the most important diplomatic gathering the world has ever known, the Third United Nations Conference on the Law of the Sea (UNCLOS III) and, in most cases, to sign also the product of that conference, which it had adopted, by 130 to 4 with 17 abstentions, on April 30 of that same year, the Law of the Sea Convention. The convention quickly attracted 122 signatures, but these did not include the United States, which had voted against adoption, or Belgium, West Germany, Italy or the UK, which had all abstained.

The convention was a comprehensive one, designed to prescribe who could do what, where, on, in and under the oceans. Thus naval activities, merchant shipping, fishing, and marine scientific research were among its concerns; but its central elements, on which the conference was to spend more time than anything else, was to be the creation of an International Sea-bed Authority, to give effect to the Declaration of Principles of the United Nations General Assembly in 1970 that the sea-bed and its resources, beyond the limits of national jurisdiction, were "the common heritage of mankind".

This ambitious undertaking involved, in effect, defining "Mankind" - devising a new body that would regulate, and tax, all mining of the international area, and, as it emerged, itself embark on such mining on mankind's behalf, and redistribute its profits and revenues for mankind's benefit. It also involved demarcating the international area, that is, specifying the criteria, and methods, by which that part of the sea-bed that fell within national jurisdiction would be distinguished from that which did not.

To arrive at a convention that would do all these things, and then adopt it by the vote of April 30th 1982, took eleven sessions of UNCLOS III, spanning 93 weeks of negotiations, spread over more than eight years.

That America and, less emphatically, several other leading mining states have, at least initially, stayed aloof from the convention is a sad disappointment. UNCLOS III had always set itself the task of reaching a convention that was not only comprehensive but generally acceptable. Hence the long years of consideration of the issue which began, not with the first session of UNCLOS III in December 1973, or even with Declaration of Principles of 1970, but in 1967, when the then Maltese Ambassador, Dr. Pardo, in a three-hour speech, introduced into the Assembly's First Committee an item entitled "Declaration and treaty concerning the reservation exclusively for peaceful purposes of the sea-bed and the ocean floors underlying the seas beyond the limits of present national jurisdiction, and the use of their resources in the interests of mankind."

Between then and the opening of UNCLOS III lay six years of consideration by, first, an "ad hoc" and then a "permanent" Sea-bed Committee, meeting twice annually and punctuated by the sessions of the General Assembly itself. Prompted and to a substantial extent educated by Dr. Pardo, the governments of the world became aware of the extent of mineral wealth that lies on or beneath the sea-bed, wealth not confined to the continental shelf, which had been the subject of a convention at UNCLOS I at Geneva in 1958, but extending, in an exploitable form, to a substantial proportion of the deepest parts of the ocean floor, to which no clear and generally accepted legal title existed.

This was not so much a problem as a stimulus and an opportunity. Could the decision as to who should exploit these new resources, and under what conditions, be settled by some more orderly process than competitive scramble, acrimonious contention, or even open conflict? and, to that end, could some new institutions be created that would manage the international area in the world's collective interest, and ensure that mankind as a whole benefited from it. The Assembly's unopposed Declaration of Principles, of 1970, in effect said yes to both questions. From then until 1973 the task of making good these commitments fell to the first of three sub-committees of the Sea-bed Committee, and thereafter to the First Committee of UNCLOS III. By 1974 even those whose scepticism about the notion of the "common heritage of mankind" had led them to abstain on the Declaration were now ready to invoke that notion as the proper benchmark by which proposals should be judged. Although UNCLOS III did not manage to convert this universally-acclaimed principle into an equally universally-accepted convention, the one it has produced already enjoys remarkably wide support and may yet form the basis of a regime for ocean mining. The primary aim of this book is to explain both the successes and the failures of this process; the secondary aim is to assess the situation in which the world now finds itself, and the chances that the "common heritage of mankind" may still become a reality.

This will involve treating this fifteen-year episode in two ways: first by delineating the framework of negotiations, and secondly by looking, one by one, at the main issues and how they were handled. The remaining chapters of the first part will therefore cover the mineral wealth .itself and the feasibility and possible economic and environmental consequences of mining it (Chapter 2); the implications of the concept of the "common heritage of mankind" from a variety of theoretical standpoints (Chapter 3); the different kinds of entities - states, groups of states and non-state actors - which participated in the process, and, in the case of the most important actors, particularly the United States, the way differing groups combined and contended to shape its policy (Chapter 4); and the choices made, both deliberately and inadvertently,in the United Nations, at UNCLOS III and elsewhere, about how these protracted negotiations should proceed, both formally and informally, and the extent to which process has influenced outcome (chapter 5).

The chapters of the second part will each consider one of the major issues that needed to be resolved. Chapter 6 will deal with the question of the geographical limits of national jurisdiction over the sea-bed, in other words the way in which the boundary between the zones of national and international jurisdiction should be drawn; chapter 7 with the system of exploitation: that is to say with the categories of entities that might exploit "the Area", the rules that should govern exploitation, and the extent to which the Sea-bed Authority should have discretion in managing it; chapter 8 with the more specific but highly contentious issue of "production policy" - how far the Authority or the convention should limit production from the Area to protect existing or potential land-based producers of the same or similar minerals; chapter 9 will deal with the structure of the Authority, that is the numbers, functions, composition, and voting requirements of its organs: and finally chapter 10 will be concerned with how disputes in this field, whether between states or between a state or company and the Authority, are to be settled.

To these five major issues UNCLOS III, in 1980, added a sixth, the question of what interim arrangements, between the adoption of the convention and its coming into force, would be made to protect existing "pioneer" investments and integrate them with the international regime, and also how the signatories of the convention would articulate, even more precisely than in that text itself, the rules of the system, and prepare for the inauguration of the Sea-bed Authority. This whole question of interim arrangements will be addressed in chapter 11. In conclusion, chapter 12 will assess the significance of the outcome of UNCLOS III for the conception of the "common heritage of mankind" and the prospects for an effective international regime.

2 The Givens: geological, technological, economic, environmental

Dr. Pardo's proposal of 1967, which set in motion this process of negotiating a sea-bed authority, was a response to the discovery of vast new mineral wealth beneath the oceans; more specifically, it seems to have been prompted by, and explicitly refers to, John Mero's study, published in 1964, "The Mineral Resources of the Sea", which had demonstrated – some might claim exaggerated – the economic potential of manganese nodules. The principle that Dr. Pardo proposed was, however, a general one. Beyond national jurisdiction, the mineral resources of the sea-bed, and not merely manganese nodules, were to be the common heritage of mankind. We must therefore consider what mineral resources there were in the sea-bed, that fell, or might have fallen, beyond the limits of national jurisdiction; what techniques have been devised for exploiting them, and at what cost, both to the miners themselves, and to the markets of the minerals concerned; and what the environmental consequences of such mining are likely to be. These are not easy questions to answer. It is not just that there are inherent difficulties, and hazards, in the calculations that their answers require, or even that such knowledge is liable to rapid and drastic change. There is the additional complication that the information we get, coming as it does largely from governments or private firms, cannot always be taken at its face value.

The sea floor can be divided roughly into two types of terrain, the continental margin, and the deep ocean floor. The latter, sometimes loosely called the abyssal plain, includes mountain ranges, isolated sea mounts, and deep trenches. The typical but by no means universal pattern for the former is for it to descend in three stages from land, the stages being continental shelf, slope and rise. The slope, markedly steeper than both what precedes and what follows it, can begin at anything between 20 and 550 metres depth, and anything between one mile and eight hundred miles from land, with an average of 132 metres and 42 miles. It usually extends from ten to twenty miles. Whether the "rise", at its foot, belongs to the margin or the deep sea-bed, is a matter of dispute, and depends on whether its surface – an "apron of debris" from the shelf and slope, – or its subsoil is stressed. As the upper layer thins and flattens out, it merges imperceptibly into the ocean floor.

It is thus almost impossible to say where the margin ends.
If we take a crude depth measure for the outer edge of the
margin, 3000 metres, we can say that an area amounting to
about 75% of the three major oceans- about half the surface
of the planet - lies beyond that outer edge.

HYDROCARBONS

Little need be said about hydrocarbons (oil and gas)
because they largely seem to be found on the shelf. As
explained below (Chapter 6), the Continental Shelf
Convention of 1958, though not universally recognised, had
confirmed a trend, already well established, for the
resources of the shelf, at least to a water depth of 200
metres, to be assigned to the coastal state.

At UNCLOS III this trend was hardly disputed, with the
exception of any part of the shelf that was more than 200
miles from land. By one estimate 728 billion barrels of oil
lay beyond the shelf, and nearly 200 billion at depths
greater than 3,000 metres. Even this, where contiguous with
shelf, slope and rise, tended at UNCLOS III to be claimed as
margin. Moreover, beyond the foot of the slope, the
occurrence of hydrocarbons can be expected to be sparser and
more expensive to work, though the development of "subsea
completion systems" and "self-positioning drill ships" seems
to have removed two major technological barriers to
increasing depth. Others remain, especially where

"local extreme weather conditions coincide with the
incidence of drifting ice floes" (Odell 1980, p.80).

On the other hand it would be premature to fix a limit, in
terms either of depth or of some other margin
characteristic, beyond which there is no oil. According to
one study

"there are indications that the presence of
petroleum source beds is very likely in the
continental slope, and progressively less beyond the
slope into the abyssal plains and ocean
depths."(Doumani 1973, p.29).

Moreover oil may be trapped in "salt-domes" in the abyssal
plains, particularly in "depositional ocean basins" such as
the Gulf of Mexico; and some have speculated that deposits
of exceptional volume might occur at the boundaries of the
vast shifting plates of the earth's outer shell. Such
speculation is however extremely precarious; and in a study
done for the UN, in 1973, was entirely discounted (UN
Secretary-General 1973).

A 200-mile limit to coastal state jurisdiction would probably have left more oil in the international area than would a 3000-metre depth limit. The UN's report of 1973 put it at 12% of total reserves (284 billion barrels out of 2272 billion). For Canada and perhaps the USA, two countries whose continental margins, in places, are particularly extensive, this proportion may be a little larger. (1)

This may become a fact of some significance, particularly since the technological and economic constraints on hydrocarbon exploitation probably increase less steeply with distance than with depth.

There can be no doubt that, in spite of temporary price reductions, the economic spur to the exploitation of sea-bed oil will become sharper. Predictions that by 2000 supplies will run out (Borgese and Ginsburg 1979, p.85) can be disregarded; there are many economic factors that could decrease demand, and increase supply, before that happens; but those that, more cautiously, predict that, by 2020, the scarcity of oil will double its price, in real terms, compared with today's (World Energy Resources 1978, p.229) are persuasive.

MANGANESE NODULES

It is however with manganese nodules that the negotiations at UNCLOS III have been almost exclusively concerned, in negotiating a regime and machinery for the international area of the sea-bed (The Area); and though much if not all of Part XI and the relevant annexes (III & IV) of the resulting convention could apply equally to other resources of the Area, practically everything in them was drafted with nodules in mind.

What are Nodules?

It is common to find all kinds of surfaces of the ocean floor, particularly the deep ocean floor, coated or impregnated with manganese oxide and/or iron oxide (Mero 1964 p.127). Nodules are accumulations of such mineral coatings about a detached nucleus (or coalescence of nuclei), which could be any hard object, most commonly "basic and acid silicates, such as pumice and glassy lapille" (p.137).

Nodules vary considerably in their composition. There are variations between one ocean and another, and between one region and another in the same ocean. There are also variations related to water depth. Even nodules from the same sampling point are not uniformly constituted.

John Mero, who first discovered the mineral potential of manganese nodules, found that nodules from the Atlantic Ocean contained more aluminium (3.1% compared with 2.9%), silicon (11.0%:9.4%), titanium (0.8%:0.67%) and iron (17.5%:14.0%) than those of the Pacific, and less manganese (16.3%:24.2%), nickel (0.42%:0.99%), copper (0.20%:0.53%), {molybdenum (0.035%:0.052%) and cobalt (0.31%:0.35%)} (Mero 1964, p.180). A later investigator, Dr. Cronan, confirmed these differences for iron, manganese, nickel, cobalt, and copper, though his averages are considerably less than Mero's for manganese (15.97%:19.75%) nickel (0.31%:0.77%) and copper (0.115%:0.3667%) (Cronan 1975, p.3834). Cronan also gives comparable figures for the Indian Ocean, where on average nodules contain an intermediate proportion of manganese (18.03%), nickel(0.51%) and copper (0.22%), but less cobalt than either (0.28%). He detects, however, a significant difference in composition between the eastern and western halves of that ocean. Beyond 70 degrees E, the average content of nickel, manganese and copper in nodules is higher, and that of cobalt lower, than in the part west of that line. Elsewhere (Cronan 1976, p.10) he includes a tract of the East Central Indian Ocean among the three most promising areas for nodule exploitation, and one of the two particularly rich in nickel and copper, although taken as a whole, even the eastern half of that ocean falls considerably short of the Pacific as a whole in nickel content. Within the Pacific, Mero delineates four compositional regions, of which only D is continuous, and that is far from compact.

TABLE 2.1
   Assays of nodules from Pacific regions on a detrital
       mineral-free basis. (Mero 1964 pp.223ff.)

	%IRON	%MAN-GANESE	%COBALT	%NICKEL	%COPPER	%LEAD
A: Along continent and between New Zealand & Tahiti	28.3	21.7	0.35	0.46	0.32	0.21
B: Near West Coast of N & S America	2.3	49.8	0.055	0.26	0.14	0.047
C: Parts furthest from land	17.7	33.3	0.39	1.52	1.13	0.18
D: Topographic High in Central Pacific	22.6	28.5	1.20	0.66	0.21	0.30
Average dry weight percentages*	14.0	24.2	0.35	0.99	0.53	0.09

*(Mero 1964, Table XXXIV, p.235)

Cronan and Tooms, (1969, Tables 4 and 5, p.340) found that nickel and copper content, and to a lesser extent that of manganese also, increases with ocean depth, while cobalt, lead, barium and vanadium content diminishes.(2)

The Pacific, and to a lesser extent the Indian, Oceans contain the most attractive regions. The Clarion-Clipperton zone, an extensive tract of the North-East Equatorial Pacific between Hawaii and Baja California, running roughly from 5 degrees to 17 degrees N, and from 160 to 110 degrees W, has generated almost all the preliminary commercial activity; but eight other large expanses, within which the nodules so far raised have averaged more than 1% combined nickel and copper content, have been identified by the US Geological Survey: five in the Pacific, two in the Indian, and one barely qualifying in the Atlantic (See Table 2.2). Even so, these promising areas outside the Clarion-Clipperton zone amount to only about 9 million sq. km, or 2.5% of the world's sea-bed, and even when that zone is added, they must fall well short of 5% of it.

John Mero's calculations of the mineral wealth, in aggregate, of the nodules of the Pacific alone conjured visions of almost unimaginable bounty: He estimated that there would be 1,600 billion tons of nodules, containing, at the rate of world consumption of each metal in 1960, 100,000 years' or more supply of manganese, nickel, cobalt, magnesium, titanium, vanadium and zirconium, more than 20,000 years of aluminium, 6,000 years of copper, 2,000 years of iron, and 1,000 years of zinc and lead. These figures, however, take no account of the practicality of bringing such nodules to the surface.

TABLE 2.2

Ocean Areas Outside the Clarion-Clipperton Zone
Containing Nodules with Average Nickel and
Copper Content Exceeding 1%
(McKelvey and Wright 1980, pp.8-15)

Locational Name for Area	Co-ordinates within which area is located (degrees)	Size of Area within which 'n + cu' content of nodules averages >1% (sq.km)	Average 'n + cu' contents within that area.
Central North Eastern Pacific	180 - 120 W, 20 - 40 N	4,900,000	1.25
Central North Equatorial Pacific	180 - 160 W, 0 - 20 N	240,000	2.00
South Eastern Equatorial Pacific	110 - 75 W, 0 - 20 S	815,000	2.11
Central South Eastern Pacific	100 - 70 W, 30 - 41 S	490,000*	1.76
Central South Pacific	115 - 119 W, 39 - 46 S	1,300,000	1.4
South Eastern Atlantic	5W- 31 E, 30 - 43 S	460,000	1.1
South Eastern Indian	95 - 105 E, 30 - 41 S	300,000	1.35
South Equatorial Indian	70 - 95 E, 10 - 20 S	270,000	2.3

* Delineation of the area within which nodules have the
qualifying metal content is particularly hazardous in this
case, where the information derives from only 5 experimental
stations.

Alan Archer (1979), for instance, defining reserves more
strictly so as to count only those of interest at current
costs and currently expected prices, holds that they must
meet three conditions: they must have combined nickel and
copper content of at least 1.76%, and an abundance, "on the
ground", of at least 5 wet kilograms per square metre, and
be located within a "prime area". (3) By this criterion,
which would disqualify half the areas listed in Table , the
oceans are thought to contain, in such "reserves", 280

million tons of nickel 230 million tons of copper, 60 million tons of cobalt, and 6,000 million tons of manganese. In the case of copper, this is rather less than the identified reserves of the metal on land; for manganese, it is rather more, and for nickel and cobalt roughly six and twenty-five times as much respectively. These figures all lie between 1% and 5% of the totals calculated by Mero, and, if account is taken of recovery and processing losses, or the tendency for high grade nodules to be less thickly congregated than low grade ones, the appropriate figure would seem to be nearer 1%. However, if prices rose, or costs proved lower than now envisaged, more nodules could be counted as reserves, and vice versa. But it is difficult to argue with Archer's conclusion that "the quantities of nickel, copper and manganese that would become available from manganese nodules seem likely to be neither enormously greater, nor enormously less, than remain to be mined on land." The case of cobalt is rather different.

Archer saw potential reserves, so defined, as covering only 0.6% of the whole ocean floor, or 2.25 million sq.km. Resources, defined as nodules with at least half the combined nickel and copper content and at least half of the abundance necessary to qualify for "reserves", (0.88% and 2.5 kilograms per square metre) are estimated to occupy about 17.5 million square kilometres, or almost eight times the area of reserves;and to amount for practical purposes (given the cumulative difficulties of lower grades, greater dispersion and disparity of terrain) to about twice reserves. This contrasts with claims by Johnson and Logue 1976(p.40, n.2), referring to an earlier paper of Archer's, for the existence of a "prime area" of around 20 million sq.km. If Archer's paper is right, manganese nodules, though clearly an important potential source (or 'reserve'!) of several metals, are not sufficiently abundant to make it unimportant who gets the first few sites (especially since the convention now speaks of 'pioneer areas' of 150,000 sq.km).

Summary

Nodules, then, containing in all very substantial quantities of manganese, nickel, copper and cobalt, to name the four most valuable constituents, cover much of the ocean floor. There is little regularity in their composition or density but those rich in nickel, copper and manganese tend to be found in the deeper parts of the ocean, far from land. The North Pacific between Mexico and Hawaii is known to be a commercially attractive region; other fairly extensive tracts in the Pacific and Indian Oceans may also prove inviting. Cobalt-rich nodules tend to be found in shallower waters; the best sites for them would probably be in the vicinity of the islands of the South Pacific and along the mountain ranges of which they form part.

There are few large regions of the ocean floor where the average combined content of nickel and copper exceeds 2%. Even those areas where this combined content tends to be above 1% amount to only a small proportion - less than 5% - of it. These relatively rich areas are likely, therefore, to be at a premium, as soon as technology has been developed to mine them at less cost than marginal mines on land, unless the pace of exploitation is regulated artificially, a possibility we shall return to later.(4)

Technological Aspects

Manganese nodules, though only recently recognised as a mineral resource, have been known for more than a century, that is, since the "Challenger" expeditions of 1873-6, (Mero 1965, p.127), to be widely distributed over the three main oceans, and have been brought to the surface in small quantities, as objects of mild scientific interest, without great difficulty. ("They did not sparkle, and they did not have a public relations agent so they were soon relegated to shelves either in the basement of the British Museum or, paradoxically, in the attic of the Natural History Museum in Washington until the discovery of their economic potential by an obscure graduate student" (i.e. Mero himself), some eighty years later. (Mero, 1975 p.343) The technical problem has been, and is, that of finding competitively economic ways of raising and processing them in bulk.

These technical problems have now apparently been solved. Northcutt Ely, for instance, who helped to prepare the claim by Deepsea Ventures to a mine-site of 60,000 square miles of the North-East Pacific in 1974, told the American Mining Congress in 1975:

"Two things do seem to be proved, however. The first is that technology now exists to harvest the nodules and raise them vertically through several miles of water to ships on the surface. The second is that technology now exists to refine the nodules and separate out the metals that they contain, ... In both respects the techniques are proprietary. They have been developed by American companies, several of which now have foreign companies as associates in one relationship or another" (Ely 1975 p.5).

The process of nodule mining can be divided, fairly clearly, into four stages: prospecting and exploration; gathering and raising the nodules; transporting them, once raised, to a processing point; and processing them, that is extracting whatever metals the miner is interested in.(5) Prospecting and exploration consist of a range of investigations, beginning at one end with widely-spaced measurement of general characteristics of the ocean floor

12

and the water column it supports, and extending to concentrated evaluations of the density and metal content of the nodules in specific areas selected as promising. At the former pole, miners need to know, in some detail, the depth of the ocean floor, the structure and depth of the sediments beneath it, and the physical and chemical characteristics of the water column it supports. They need this information both to identify where the richest deposits are likely to be found, and to decide what equipment they will need to design, and how they can best use it, to exploit them. They are thus the avid consumers of certain general kinds of scientific knowledge about the oceans, some which in the last resort they may have to produce privately for themselves.

Water depths can now be determined by echo sounders; and seismic devices such as pneumatic sound generators can unearth much material about the structure of the sea-bed, some of which might prove helpful for other kinds of mining. Sensors attached to cables can record a variety of characteristics of the sea-water; and all this information can be augmented by satellite photography. At the narrower and more specific end of the scale, still and TV cameras, now capable of being protected against the disturbances from either the ocean floor or the movement of their ship by means of elaborately-devised depressor platforms, can go far to determining, within quite short distances, nodule concentration. The metal content can be ascertained only by the taking of samples. This can be done by a number of devices, two of the most notable of which are "freefall grab samplers" and spade corers. The grab samplers have a buoyant core which is then weighted with ballast so that they sink to the sea floor. When they reach it they jettison the ballast, close their jaws, and rise to the surface. Spade corers permit estimation of both metal content and concentration, at a given point, by extracting a sample (of up to 45 cm. length) of sediment together with the nodules lying on it undisturbed (Metallgesellschaft 1975, pp.18-26).

A variety of methods have been devised for gathering and raising the nodules from the ocean floor to the surface. Only one is mechanical, the continuous line bucket system (CLB); the others are essentially hydraulic, either in whole or part.

The CLB is the simplest to understand. It consists of a cable loop with buckets attached, long enough to extend from the ship to the ocean floor 5,000 metres below and continuously rotating so that each bucket in turn travels to the ocean bed, scoops a mixture of nodules and sediment from its surface, and is then lifted back to the ship. Tests with a system of this kind in Hawaii in 1972 were encouraging, and several improvements were introduced, including the use of two ships instead of one (thus requiring the empty buckets to return after unloading across

13

the stretch between the ships to complete the loop).

Other systems involve some form of hydraulic lift, or air-lift, combined with a dredge. In its simpler form, the dredge itself has a suction pipe at its rear, so that the material is sucked off the sea-bed as if by a vacuum cleaner. Deepsea Ventures developed a nodule collector of this kind, and tested it in 1974, in 800 metres of water on Blake Plateau. The dredge and suction pipe are suspended on springs at three points within a sledge on runners to enable them to negotiate difficult terrain. The suction pipe in this case operates by airlift hydraulic dredgers to bring the nodule and sediment mixture to the surface (Metallgesellschaft 1975, p.27 and Figure 23 on p.29). Another device produced by DEMAG AG (Metallgesellschaft 1975, p.28 and Figure 24) consists of a case, 10 to 15 metres long, mounted on caterpillar tracks, with a moveable suction pipe at the front.

One problem with all these devices is that if they are to achieve an output regarded as economic, either the dredge must be quite wide or they must move fast. For instance, as Gauthier and Marvaldi point out, to produce 10,000 tonnes of wet nodules a day from a deposit with a concentration of 10 kg. per sq. metre requires a dredger to advance over an area at least 24 metres wide at a speed of 0.5 metres per second; or if the width is reduced, at a correspondingly faster speed. (Gauthier and Marvaldi 1975, p.346) Moreover, that assumes the dredger raises all the nodules it encounters in the swathe it carves in the sea-bed, that is a sweep efficiency of 100%. As sweep efficiency falls, either the width or the speed has to be proportionately increased. Thus Metallgesellschaft conclude that the DEMAG AG device will probably not be economic, if directly attached to a suction pipe. With a pipe of 15 metres long, it can only, at most, cover a band of 30 metres, plus twice the distance, (which may well be negligible) that its suction power extends beyond its head.

For this reason, and because a continuous system can stand up to what may be expected to be extreme working conditions only at very low speeds, it has been suggested that it might be more efficient to separate the process of collection from that of raising the nodules to the surface. Thus the DEMAG AG collecting device might be used to set up tracts of high nodule thickness - say 0.2 tons per metre, which could be picked by a conveyor pipe (Metallgesellschaft 1975, p.28). Another development that would permit this is the drag-net, rather similar to a fishing trawl, which would gather and sieve through the nodules, leaving behind it a concentrated trail of medium-sized specimens to be raised to the surface. Separating the two processes permits the speed of the collection device to be regulated independently of that of the entire system; more than one device could be used to service one pipe; and perhaps some of the waste and disturbance to the water column involved in transporting to

the surface a mass of unwanted sediment two and a half times
the mass of the nodules could be avoided by the sifting
process on the ocean floor. At concentrations of 10 kg. per
sq.metre, a hydraulic system might produce 5,000 tons a day
(as against 3,500 tons for a CLB system), whereas at
concentrations of only 5 kg. per sq.metre, the CLB system,
by enlarging its daily area, could still raise 3,500 tons a
day while the daily output of a continuous hydraulic system
would fall to 2,500 tons (Gauthier and Marvaldi 1975,
p.349). However, the daily yield of a system which
separated the collection from the raising of the nodules
would not depend so rigidly on nodule concentration.

The second measure of the efficiency of a nodule mining
operation, the proportion of the nodule population of a
given area that will be raised, is in no case expected to
exceed 50%. The chief scientist of what was then Deepsea
Ventures, W. D. Siapno, in a paper to the American Mining
Congress, calculated that in order to achieve an output of
1,500,000 tons of wet nodules a year for twenty years from a
site with a concentration of 1lb. per sq.ft., (roughly 5,000
tons per sq.km), the site would need to have an area of
between 23,000 and 54,000 sq. kms., in other words, a site
containing between 115 and 270 million tons of nodules would
be needed in order to harvest a total of 30 million nodules.
This would mean that between 10% and 25% of the weight of
nodules existing in a given site would be recovered.

Four assumptions underlie this calculation. First, it was
assumed that 10% of the area in a site would be unmineable
because of roughness of the terrain or other obstructions;
secondly, it was assumed that a further 10% would not be
mined because the nodules it contained would not be rich
enough in metal content; thirdly, in the remaining area,
only 45% was expected to be actually swept by the
dredgehead, as a result of the "combined effects of
currents, winds, swells and waves" and "in deference to
ecological considerations", which "demand a major percentage
be left undisturbed". Finally, in the area actually dredged
- amounting to 36% of the original site - the dredge
efficiency was anticipated as between 30% and 70%, that is
between 30% and 70% of the nodules "encountered" would
actually be lifted (Siapno 1975, pp.35-36).

Comparable estimates from other sources make these figures
look cautious. Gauthier and Marvaldi (1975 p.348) spoke of
the CLB system achieving a "nodule collecting efficiency"
(corresponding to Deepsea Ventures' "dredge efficiency") of
80%. David B. Johnson and Dennis E. Logue 1976, citing a
paper by Alan Archer written in 1974, assume that, overall,
between 20% and 25% of the nodules in a site would be
gathered up by a first-generation mine. The
Metallgesellschaft AG Review speaks of a "recovery rate" of
25% to 50%, without at that point specifying the kind of
system to which it relates (1975 p.18). Post (1983, p.26)
gives 15%-20%. Even so, and even allowing for the

restraints imposed by ecological considerations, present techniques of nodule mining, particularly that proposed by Dr. Siapno, are likely to prove wasteful of site space. The hydrolift system, though, is said to be more efficient than the airlift system. Its chief disadvantage vis-a-vis the latter is that in it the moving parts which create the lift operate at great depths, whereas in the airlift system, the lift is produced by an air compressor located on the ship (United Nations 1980, p.2).

Once the nodules have been raised to the surface, they will need to be transported to a shore plant, unless they are being processed at sea. The latter would enormously reduce the bulk and weight of the cargo to be shipped, especially if manganese were not being extracted, (instead of 3,000,000 tons of dry nodules, say, a year, it would be a mere 80,000 tons of metal precipitate (Wright 1976, Appendix A, p.2)), but capital and energy costs and (much less importantly) labour costs would be higher, certain processes (such as electroplating nickel) might still require a plant on land, so that even apart from environmental considerations, to be discussed below, it may not be attractive.

In any case, it is technically quite possible to organise processing on land. The mining station which receives the nodules as they are lifted to the surface will not itself transport them to the shore, and the chief problem lies in transferring them to the carrier ship. The mining station has to be held on course, regardless of sea and weather conditions, and the carrier will have to be physically coupled to it, an operation which in severe conditions would need to be finely judged to avoid a collision or a breakdown in transshipment. The Review by the Metallgesellschaft AG describes a process called Marconaflo for the storage, transshipment and transport of crushed solid minerals in the form of a slurry. This involves grinding down the nodules and mixing them with water on the mining station so that they can be pumped through a pipe to the transport ship. To reap the economies of scale the latter would need to be large (70,000 tons dead weight), with shallow draft (10 metres) to avoid limitation as to port of call, and equipped with strong transverse thruster plants with direct current propeller motors in the fore and aft to enable it to come near enough to the mining station and stand off it without being towed or supported by the latter. It is thus no ordinary ship, but certainly such ships can now be built, if indeed they have not been already. (6)

These figures, however, relate to only one link in the processing chain. Whatever similarities such links, taken in isolation, may have to stages in the refining of ores mined from land, the chain as a whole, for each firm concerned, has been devised specifically for processing nodules. As Dr. Kruger and Dr. Schwarz put it,

"It goes without saying that none of these processes (i.e. those now used in treating nickel ores) can be directly transferred to the processing of manganese nodules" (Kruger and Schwarz 1975, p.37).

Perhaps; though the case has not been proved. The argument is that, as will be explained later for environmental reasons, hydrometallurgy is preferable to pyrometallurgy, a point equally applicable, on the face of it, to mining on land; and that the composition of the nodules themselves, including their secondary components and impurities, require a tailor-made processing operation. It would be overstating this case to claim, as Ely and Flipse do, (7) that processing plants must be designed anew for every deposit. Since mineral content of nodules varies sharply even at the same point, any processing plant must be fairly versatile. Different processing plants for different sites could only make sense in cases where composition varied more between sites than within them.

What is undisputed is the fact that the processing and refining stages are not likely to present any formidable impediment to completing the chain of operations of nodule mining floor from the ocean floor to the processing and refining stage. (8) As the two contributors to the Metallgesellschaft Review put it,

"The main problem in this connection will not be concerned with the individual process steps but rather with the necessity of developing a comprehensive solution for a process which fully meets the requirements of technological feasibility, reasonable production size and environmental control within an overall economic concept." (Kruger and Schwarz 1975, p.43).

Thus Northcutt Ely's claim that the technology now exists to locate, gather, raise, transport and process the nodules seems well substantiated. That it is, or will remain, the trade secret of a few American companies and the consortia to which they belong is less credible. Particular designs may remain proprietary, but the essentials of the techniques necessary to exploit this resource seem to be sufficiently well known to allow any entity with enough money to enter the field.

Exactly how much money will be required to do this is not easy to assess.

Many figures have been put forward, and the variation among them, even after allowing for the fact that they may be expressed in terms of dollar values of different years, is striking. Table 2 illustrates the extent of this variation. It presents six early estimates of the capital costs and working costs of nodule mining operations. In all but one case revenues are also estimated, which permits the profit, per ton of nodules, and the rate of return on capital, to be calculated.

These estimates now seem to have been far outdistanced. By late 1981 Marne Dubs, speaking on behalf of the Kennecott Group, mentioned

"capital investments of the order of the one and a half billion dollars ($1500 million) required for a commercial-scale mining operation". (15)

The enormous discrepancy between figures like these, and John Mero's claim to be able to embark on sea-bed mining with capital of a mere $10 million, is puzzling. Mero's operation would have used a CLB system, used sites of only 5,000 to 6,000 sq.km, and extracted only nickel and copper, expecting to recover 60%-70% of the content of these metals. The UN Report of 1980 does not mention his company, Ocean Resources; and it seems that CLB systems have come to be regarded as impracticable. Whether this was the only impracticality in Mero's scheme is not clear.

18

TABLE 2.3

A Comparison of Some Early Estimates of Costs, Revenues and
Pre-tax Profitability of Manganese Nodule Mining Operations.

SOURCE	Scale (m. tons)	No. of Metals	Base Year for Prices	Capi/tal inc. Work/ing $m	Oper/ative Costs per ton $	Revenues per ton $	Profits per ton $	Profits per yr. $m.	Gross Rate of re/turn on Cap PreTax %
UN 1974 (9)	3	3	n.s.	250-280	20-30	81-99	51-79	152-236	54-94
	1	4	n.s.	154-188	56-76	154-188	78-132	78-132	43-109
Johnson & Logue (10)	3	3-4	1970	150-300	13-24				
Metall Gesell/schaft (11)	3	4	n.s.	425-475	40-65	100-125	35-85	105-255	18-50
Mero (12)	0.3	2	n.s.	10	20	50	30	9	90
Wright (13)	3	3	1975	500-750	40-55	93-103	48-63	138-183	18-37
Leipz/iger & Mudge (14)	3	3	1974	300	29-40	98-138	58-109	174-226	45-87
-"-	1	7	1974	140-210	74-99	284-325	185-253	185-253	82-112

The markets for nodule metals.

At most, prospective nodule miners have spoken of
producing seven metals from nodules: cobalt, copper,
manganese, molybdenum, nickel, vanadium, and zinc. Three of
these (molybdenum, vanadium, and zinc) were mentioned only
in connection with what was then Deepsea Ventures (Leipziger
and Mudge 1976, pp.147 ff.) and do not appear at all in the
1980 UN Report, so attention here will be concentrated on

the other four. No attempt will be made to predict the future of these markets. A recent study of past predictions (Page and Rush 1978) has shown how precarious they tend to be and how quickly the assumptions on which they rest become out-of-date.

Copper. Copper is the largest of these markets. World primary consumption in 1976 was estimated as "close to 8 million tons" (US Congress 1978, p.48). A nodule operation on the scale we have been discussing (3 million tons of dry nodules per year) would therefore meet something like one two-hundred and fiftieth (i.e. 0.4%) of the world's present demand. Growth of demand in the long run has been fairly steady, at between 4% and 5%, since 1900 (Prain 1975, p.43). Demand seems to be price- elastic, at least in the long run (Leipziger and Mudge 1976, p.171 n.13; Johnson and Logue 1976, p.51); it is a metal "vulnerable to substitution", especially by aluminium (US Congress 1978, p.47)(and, presumably, capable of being a substitute for this and other metals in many uses).

The 1980 UN Report, while noting that major producers of copper could be adversely affected by nodule production, does not even list its major producers, judging that "the impact on the copper market should be minimal" (UN 1980, p.3). This is perhaps an exaggeration. The size of the copper market, and its fairly steady growth, and fairly high price-elasticity of demand, make it likely that copper output from nodules will, for the next twenty-five years constitute a not very substantial proportion of the market's growth segment - say not more than 20% at most, and that the addition it makes to what is otherwise available will be absorbed without the price being much lower than it would otherwise have been. As a result, the effects of sea-bed mining will be swamped by other factors affecting the copper market, and very difficult to isolate. Nevertheless, a small, even a very small, change in a large market may not be negligible, and although the main exporters include developed states (Canada, South Africa, and Australia), developing countries account for about half world production outside the USA (a net importer). For six developing countries, copper exports, in 1974, constituted more than 10% of all exports (Zambia 93%; Zaire 73%; Chile 67%; Peru 23%,; Philippines 14%; Cyprus 11%) (Leipziger and Mudge 1976, Table 5.3, p.134).

So far, for simplicity, it has been assumed that the copper market is a competitive one. It has been argued that competitive markets tend to be highly volatile, and that copper, like lead and zinc, has been "subject to tacit collusion by major international corporations" (Krasner 1977, p.47).(16) Even so, its price appears to have been quite volatile, with year-to-year price changes averaging 13.6% in the quarter-century up to 1974 (though this figure, by disregarding long-term contracts, "probably exaggerates" its volatility) (Krasner 1977, p.46). Insofar as copper is a managed market, and insofar as the major corporations that manage it are also involved in sea-bed mining, it may be that they will effectively insulate the market from sea-bed competition.

Nickel. Nickel is a much smaller market than copper, total output being in the region of 625,000 tons in th 1970's (666,000 tons in 1976). Until recently it had been distinctly more buoyant, world demand having increased by an average of 6.3% per year between 1890 and 1973 (Archer 1973, p.I-318). This annual rate of increase slowed down markedly in the nineteen seventies and estimates of growth rates of less than 3% p.a. for the rest of the century began to be not uncommon (US Bureau of Mines 1975; Malenbaum 1977).

Within UNCLOS III's own Negotiating Group I (whose mandate included the question of production limitation), a sub-group of technical experts, under the chairmanship of Alan Archer, discovered a fairly abrupt change in 1967, in the rate of increase of mine production, refined production, and consumption of nickel. In consumption, in the previous ten years the average rate of increase had been over 9%; in the ten years that followed, less than 4%. This led the sub-group to predict, cautiously, average annual growth rates lying between 2% and 6% between 1977 and 2000 (UNCLOS III, Records, Vol.X 1978, p.45).

According to the UN's 1980 Report, 34% of 1978 mine production, and 61% of ore reserves, were in developing countries (including in each case Cuba); 42% in developed market economies; and 23% in 'centrally planned economies'. The largest producers are Canada (30%), the USSR (17%), New Caledonia (15%) and Australia (11%). This represents a sharp increase in the proportion derived from developing countries, which in 1974, which had been little more than 10%. The Dominican Republic and Indonesia are, outside Cuba, the largest developing-country producers (Leipziger and Mudge 1976, p.141).

A single nodule-mining operation of 3 million tons a year could produce something approaching a twentieth of current world output of nickel, so that ocean mining could account for a substantial fraction of that output by 2000, particularly if demand grows slowly. Its price, though not by any means inelastic, (17) is probably less elastic than that of copper. On land sulphide ores, which make up the bulk of existing deposits, are tending to run out, and new additions to reserves have been in the form of lateritic ores. Apparently sea-bed mining is competitive with the latter but not with the former (Wright 1976, p.7). Lateritic ore producers therefore have some cause to be concerned with the effects on their income of sea-bed production.

A glut in 1980, with 35,000 tons added to world reserves and consumption falling to 535,000 tons (Post 1983, p.45), would suggest that insofar as nickel is the prime reason for ocean mining, the latter will not take place unless subsidised. Nickel demand is seen as particularly sensitive to the overall buoyancy of the free market economy.

Cobalt. The cobalt market is a small one. World mine production, in 1972, was only 23,000 tons, less than 4%, by weight, of that of nickel (UN 1974).

Assuming a cobalt content of 0.2%, it would take only four major mine-sites, lifting 3 million tons of nodules a year each, to produce the 1972 world output of cobalt.

Present production, and land-based reserves, of cobalt are more heavily concentrated in developing countries than is the case with the other three main nodule metals. In 1978 such countries accounted for 67% of production and 72% of reserves, Zaire contributing more than a third of world production, New Caledonia 14% and Zambia 7%. The main developed country producers were Australia (11%) and Canada; in all such countries have one-fifth of world production (UN 1980, p.4).

Demand for cobalt is closely related to the general level of industrial activity. Income elasticity is about unity: that is, if income rises by 5%, cobalt demand will rise by 5%, and so on. In its main uses, (18) a given rise or fall in price will affect demand relatively little, except that, since cobalt is a good substitute for most uses of nickel, its price is unlikely to fall below the latter's. In recent years, however, the price of cobalt has soared far above that of nickel. From $6 a lb. in 1977 it rose to $60 but fell again, by March 1981, to $20. This volatility was mainly due to interruptions to, and uncertainties about, supply, particularly in the wake of disturbances in Zaire; but the increasingly exacting demands for purity - for some currents needs it has to be 99.7% - has apparently reduced the usefulness of the US stockpile (Post 1983, pp.47-48).

Since uses of cobalt, other than those in which it is a substitute for nickel, cannot be easily expanded, and since on any substantial scale nodule mining is likely to produce several times the present world output of it, if there is ocean mining the price is likely to fall to little above that of nickel. At the time of Caracas the difference between these two prices was small enough to make it feasible to talk of compensating developing-country producers As the gap widened, the cost of this option began to look prohibitive.

Recent cobalt prices, though much higher than a few years ago, should not greatly affect decisions to invest in sea-bed mining, since these have to be taken at least five years in advance of the first commercial production, and there can be little assurance that prices in 1988 will bear any relationship to prices in 1981 or 1983. Moreover, since the market is so small, the price is not a "given" but is quickly capable of being brought down by output of nodules.

Manganese. Although, in nodules, manganese is vastly more abundant than the other metals whose extraction is being contemplated, amounting to between 20% and 50% of nodule weight (see above Table 2.1), there was considerable uncertainty, at first, as to whether nodule-miners would produce it. When Leipziger and Mudge were writing, only Deepsea Ventures, later to become Ocean Mining Associates, were proposing to do so, and then were thinking in terms of raising only one million tons of nodules a year. The UN Report of 1980 includes manganese among the four metals but gives no details of how many consortia propose to extract it.

There appear to be two fairly distinct markets for manganese - a small one for manganese metal and a much larger one for ferro-manganese, and manganese ore. The manganese metal market amounted, in 1972, to only a quarter of a million tons, which could be entirely met by an operation of the scale Deepsea Ventures had in mind. The metal content of world production of manganese ore in 1971 was 8,300,00 tons. As now consumed, however, such ore has a minimum metal content of 35% which is higher than that found in most nodule samples. The market for ferro-manganese was comparable in size (between five and six million tons) but has a much higher metal content (between 74% and 95%). In 1974 the prices of ferro-manganese and manganese metal were respectively, six and eight times that of manganese ore (31c and 42c per lb, compared with 5c). Therefore nodule manganese would need some refining to compete even with manganese ore, and much more to compete with ferro-manganese.

The rate of growth of demand for manganese seems to have followed a similar path to those of other nodule minerals. At Caracas, there were great expectations based on past performance (5% per year). By the late seventies, in the light of the slowdown of the world economy, these estimates were halved.

Developing countries produce a smaller share of world output of manganese than they do of the other three main nodule metals (27% in 1978) and account for an even smaller proportion of its reserves (7%) (UN 1980, p.4). Gabon and India each produce 8% of world output, but only in the former case does this item account for a significant fraction of the country's exports (about 10%). Other developing-country exporters in 1974 were Brazil (the largest, in 1970 and 1971), Ghana, Zaire and Morocco. All these are dwarfed by the two leading producers, the USSR (38%) and South Africa (23%), who also process, between them, a preponderance of the reserves (87%) (Post 1983, pp.49-50). There is some plausibility, therefore, in the hypothesis that Western interest in the extraction of manganese from nodules is partly based from a judgement as to the political undesirability of becoming dependent on either or both of these two main sources of manganese, and not just on the expected profitability of the process, in the present and foreseeable state of the market.

Economically, if manganese is taken from nodules it will flood the market for manganese metal. If nodules can also be used to produce ferro-manganese (or if manganese metal is a good substitute for it), the price of the two will fall together. Demand for manganese appears to be inelastic; there are few close substitutes, and few alternative uses. However, since manganese exports from developing countries are relatively small, compensation for export losses might be more feasible than in the case of cobalt.

Vanadium, Molybdenum and Zinc. Vanadium is a small market which nevertheless expanded rapidly between 1964 and 1974 (from 9,000 to 28,000 tons) and is expected to enjoy rather brighter prospects than most metals for the remainder of the century (US Bureau of Mines 1975, pp.1206-1207). It forms about one-tenth of one per cent of nodules. Like molybdenum and zinc, it has been mentioned only in connection with the smaller-scale Deepsea Ventures operation, whose annual output of one million tons of nodules would supply about 3% of 1973-76 world output, 1% of anticipated world output by 2000. If other miners extracted it, it would not take many operations to make a large dent in the market, especially if the market's growth is less than anticipated. As with manganese, the USSR and South Africa are the chief exporters; they have within their borders 90% of the world's reserves.

The zinc market is large (around 5 million tons in 1973-5) and zinc makes up less than one tenth of one per cent of nodules. Any zinc produced from nodules is not then likely to affect either the world market for the metal or the profitability of an operation.

Molybdenum, by contrast, is a small market and about the only nodule metal of which the United States is a net exporter. World mine production of it doubled between 1964 and 1974 to reach a level of between 80,000 and 90,000 tons a year in the 1970's, and a further tripling was predicted by 2000. Its price fell in real terms between 1964 and 1973, after rising in both real and money terms for the previous decade. It recovered, however, in 1974 and 1975 to reach between $2 and $3 a lb.

Nodules contain rather more than 0.1% of molybdenum, and although this proportion is low, it is not much lower than in some ores from which the metal is currently produced, mainly as a co-product with copper. A million tons of nodules could yield about 1,500 tons of molybdenum, about 2% of current world output and rather less than 1% of projected world demand by 2000. If most ocean miners extract it, and if the buoyant estimates of demand growth made by the Bureau of Mines prove excessively optimistic, nodule mining could make a noticeable impact on the world market before the end of the century.

Environmental Aspects

The mining of manganese nodules could have worrying environmental consequences, particularly if they are processed at sea. Life in the vicinity of the deep ocean floor far from land is sparse but vulnerable. (19) If whole species are not to be unthinkingly exterminated, it will be necessary to alternate between mined and unmined strips (as the technique of Dr. Siapno, of the former Deepsea Ventures, described above, envisages), though even that could cause a turbidity that might prove fatal to "filter feeders" (sestonophages) including sponges, or could alter the chemistry of the bottom layer. A CLB system for raising nodules would also produce turbidity throughout its traverse of the water column; and would raise much unwanted sediment with the nodules. To dispose of this without increasing the turbidity of the water column would require the use of a second discharge pipe reaching back to the sea floor. Effluents derived from washing the nodules discharged at the surface could have unpredictable effects on the biological systems of the water-layers through which they pass, introducing new organisms, stimulating others, reducing the light absorbed (near the surface), and accumulating toxic metals, like cadmium, in the "food webs".

These are possibilities, not probabilities. They are arguments for requiring that nodule mining proceeds fairly slowly, and carefully, and is thoroughly monitored, so that, if such possibilities materialise, prompt counteraction can be taken. Processing at sea would give rise to hazards of quite a different order since its raison d'être would be to reduce the load to be transported to land so that the latter would consist merely of the refined metals thereby produced. If manganese was not being extracted, this would mean emptying 95% of the weight of the nodules back into the sea. If it were, the figure would be about 70%. Thus each three million tons a year operation would entail dumping between two and three million tons of nodule waste a year in the ocean, plus the chemicals used to refine them, many highly pollutive, with heavy alkaline or acid bases. (Glasby 1977, p.436) Again, we do not know what the environmental consequences of this could be. The waters in which nodules are likely to be mined are not, as far as we know, rich in fish or tourist potential, though they have westward-flowing currents moving at between 1,000 and 2,000 miles a year which could spread any poisonous substances, possibly in a patchy way; fish and other living organisms, not themselves vulnerable to these poisons, could also spread them through a food chain that ends with human beings.

If, on the other hand, the nodules are processed ashore, nodule mining might well be environmentally less damaging than land-based mining of the same metals. Copper is now being derived from ores containing only 0.5% of the metal, thus generating twice as much residue, per ton of refined copper, than nodules with something around 1% of copper content; add the wastes from producing on land the same quantities of nickel and cobalt, and the advantage of nodule-mining becomes even more pronounced. Moreover, the use of hydrometallurgical rather than pyrometallurgical techniques (leaching instead of smelting) is also an environmental improvement. It would, however, be premature to conclude that because the mining takes place in a remote and normally invisible theatre, it is necessarily less devastating than mining on land. The rapidity with which we can observe environmental consequences is an important factor in counteracting them. Society will not discover the environmental consequences of sea-bed mining unless it takes specific steps to do so. There is little reason to think that nodule mining will bring ecological catastrophe, but to ensure that it does not, global environmental standards need to be applied, and constantly revised, in the light of new understanding of the effects of such mining, derived from close monitoring of its impact.

Given that these wastes have to be disposed of somehow, the advantage of disposal on land is that the hazards, and disfigurement of the environment, can be localised, and regularly checked. Disposal at sea has the advantage that, up to a point, toxic minerals can be absorbed and rendered harmless, but it is very difficult to know where that point is. Thus Annex I of the 1972 London Convention on the Prevention of Marine Pollution by Dumping of Wastes and Other Matter has listed substances which cannot now safely be dumped in the sea at all, as well as others whose dumping needs to be regulated. It seems that for such substances the sea is approaching, in at least some regions, the limits of its absorptive capacity. The sustained dumping, on a very great scale, of the residues from processing nodules at sea would introduce such risks of long-term or even permanent environmental damage that the prudent should rule it out of court.

POLYMETALLIC SULPHIDES

The metalliferous muds of the Red Sea have been known for some time. Discharges of hot water, rich in metal, form pools whose sediments contain significant quantities of zinc, copper, lead, silver and gold (Bischoff and Manheim 1969, pp.535-541, cited in Ross 1978, Table 2, p67). The Red Sea is a boundary between two of the Earth's plates, and a "zone of divergence" where these are pulling against each other. It will also fall entirely within national jurisdiction for resource purposes, that is within the economic zones of the riparian states. Until recently, it was not known whether similar concentrations of metals could be found at other "zones of divergence" between plates, many of which run through the open ocean. It was argued that

"This accumulation aspect appears to be unique to the Red Sea and results from the fact that the sediments buried beneath the flank of the sea contain salt deposits that have been leached by the migrating fluids and thus the fluids have increased in density" (Ross 1978, p.68).

In August 1981, what appears to be a similar phenomenon was found in the Galapagos rift in the Eastern Pacific, 240 miles east of the Galapagos Islands and 350 miles west of Ecuador. Here, however, the thermal vents from the rift valley appear to form "chimneys" as they enter the near freezing waters of the abyssal depths. The average metal content of samples from this deposit is: 10% copper; 0.03% silver; 0.01% cadmium; 10% iron; 0.1% molybdenum; 0.1% lead; 0.03% tin; 0.1% vanadium; and 0.1% zinc. The East Pacific Rise and the Juan de Fuca Ridge have also been found to contain sulphides rich in metal, particularly zinc, which appears to constitute about 50% of the deposit, and in the former case, copper (up to 6%). These finds seem to confirm the speculation that those parts of the hinges of the earth's plates that fall outside national jurisdiction may constitute another substantial asset of the "common heritage", and moreover, one that may not be depleted by exploitation, since the Galapagos chimneys are apparently growing at a substantial rate (a foot in two years), which would permit them to be "harvested" at intervals.

It is not yet clear how these will be mined, though no insuperable difficulties are foreseen. Their main commercial attraction seems to lie in their zinc and copper, and perhaps also silver. If they were to be located at a large number of points on these plate boundaries they could have a severe effect on the market for these three metals.

The environmental impact of such mining could also be considerable. Where the thermal vents are warm rather than hot they produce an array of fauna quite unlike anything else on earth, and a manned expedition to the Galapagos Trench in 1977 revealed a rich and scientifically sensational assortment of species inhabiting a tiny area of

27

deep sea floor in the vicinity of the diving chamber. Since it is not yet known what techniques would be used in mining these sulphides, the extent to which it would pose a threat to the survival of such creatures cannot be assessed; but it certainly cannot be assumed that, without conscious precautions, mining and environmental protection are compatible.

More potential mineral resources of the sea-bed could have been listed. Some, like sand and gravel, are for obvious reasons confined, as exploitable assets, to parts of the sea-bed that would on any definition of it have fallen within coastal state jurisdiction. Others are highly speculative - as were polymetallic sulphides in the open sea until 1981. It is clear, though, that there are substantial and important resources which, on any definition of the sea-bed beyond national jurisdiction, will in large part come within it and thus give economic substance to the notion of the "common heritage", and that the economic and environmental consequences of exploiting them demand international regulation.

NOTES

1. Figures vary for the USA. Robert Hodgson, the Geographer in the US Department of State, puts it at no more than 5% at most (Hodgson 1975, p.190). Another source estimates that 14% of US oil reserves and 12% of her gas reserves lie in water between 2500 and 4000 metres deep and a similar proportion — slightly larger for gas — further than 200 miles from her coast (Johnson and Logue 1976, n.19 on p.57). Don Sherwin, of the Canadian Dept. of Energy, commenting on Hodgson's paper, put the Canadian figures at 16% (Hodgson 1975, p.197).

2. Copper from 0.1% (less than 3000m), to 0.5% (5000+m); nickel from 0.3%(<4000m) to 0.6%(>4000m) cobalt from 1.3 (at 1000 metres) to 0.3%(>3000m).

3. Thus he ignores any apparently isolated pockets of nodules, even of satisfactory grade and local abundance, assuming that profitable mining would require concentration on a given area, or "site". He also assumed that "metal content" and "abundance" were independent variables, whereas there is now some evidence that they are negatively correlated, which would further reduce the figures for reserves. On the latter point, see Menard and Frazer 1978.

4. See chapter 8 below.

5. There has also been some talk of processing the nodules at sea, which would make transport (of the refined metals) the final stage, and a much less costly operation than transporting the nodules. But it would probably create some environmental hazards, in disposing of the wastes, and

require energy sources to be transported from land to the processing ship. The environmental question is discussed later in this chapter.

6. This account is largely derived from Reidel 1975, pp.34-35. John Mero, who has concerned himself with developing and marketing the technology for small-scale mining, envisages ore carriers that are "assumed to be barges or small 10,000 tons or so vessels", but presumably the problems here would remain (Mero 1975, p.327).

7. "You must design both your mining system and your processing system to suit the deposit" (John Flipse 1975, p.329). Flipse also argues here that "if we had to go from one deposit of nodules to another deposit, it could cost us anywhere from 30 to 75 per cent of our capital investment". Similarly Northcutt Ely, like Flipse associated with what was Deepsea Ventures, contends that "the metallurgical problems involved in refining nodules of different deposits located only a few hundred miles apart may differ so completely as to render the refining process that is developed for the one uneconomical for the other (Ely 1975, p.6). It should be borne in mind that, if such firms ever have to pay, either to an international authority or a national government, taxes based on profits, the nominal profit will be that much less if the processing plant (and other investment) for working a given site can be "written off" and thus count entirely as costs than if it is deemed to have produced an asset that remains valuable after the site had been fully worked.

8. Though it is reputed to be expensive. Hasegawa (1978, p.81) sees processing as consuming 65% of the "operation cost" of nodule mining.

9. UN 1974 ("Economic Implications of Sea-bed Mineral Development in the International Area: Report of the Secretary-General"). The figures for operating costs are derived from answers reportedly given by Marne Dubs, of Kennecott, to questions following his paper to the 9th Annual Conference of the Marine Technology Society, Washington, D.C., September 10-12 1973. Which end of the range applies appears to depend on "whether or not strict environmental regulations were enforced" on processing, but they are in any case much lower than an estimate, also produced in 1973, by A.J. Rothstein and R. Kaufman. See also p.68 of the UN Report and Table 7 on p.69.

10. Johnson and Logue 1976, p.44. Johnson and Logue cite "Economic Implications" above as one of the two sources for their estimates.

11. Boin and Mueller 1975; the capital investment figures are taken from the graph on p.49 and the operating costs and gross revenues from Table 6 on p.48, with DM converted in dollars at a rate of $1 = 2.40 marks. The figures are

subsequently rounded.

12. Mero 1975, p.344 (for scale and capital costs), pp.345-6 (for operating costs and revenues).

13. Wright 1976, Table 6, p.11 (for costs, "Working Capital" has been added to "Total Investment" to give capital costs in my table). Revenues are derived from Tables 7-10 and allow for falls in the price of nickel and cobalt when seabed production has become established.

14. Leipziger and Mudge 1976, ch.6. The operating costs (in which debt amortization is included) and total revenues per million tons for the two types of operation are given in Table 6-4, p.160. The investment costs are discussed on pp.158-160. They include the cost of foregoing alternative uses of capital during the interval between outlay and revenue intake.

15. "Questions-and-answer session before the US Congress", October 1981, reported in Neptune, No. 19, March 1982, p.8.

16. A contrary view is put in Harris 1980, p.202.

17. "In essentially all its uses, other materials can be substituted for nickel. However, in most cases substitutes would entail increased cost reduction in some specific property or characteristic which would, consequently, affect the economies or performance of the product". (US Congress 1978, p.42). The substitutes mentioned are columbium, molybdenum, chromium, cobalt, vanadium, manganese, copper, platinum, aluminum, titanium and (for coating) plastic paint or enamel.

18. High temperature alloys, permanent magnets in loudspeakers, motors and generators, in steel for high speed cutting tools, as a dryer in paint varnishes and electroplating, and in the desulphurisation of petroleum and coal, including possibly car exhausts (See Adams 1973, pp.2-5).

3 The Notion of the Common Heritage: Implications and Critique

Pardo's phrase, "the common heritage of mankind", has always had a ring to it. From the start, it has meant more than a global commons, open to all to graze on. It has implied the establishment of rules by which the exploitation of a part of the earth's resources are to be governed; and of institutions capable of acting on behalf of mankind as a whole. It therefore raises the question: is this a function with which it is feasible to endow any organisation?

There are several schools of thought in international relations that would have to answer "No". The "realists", for instance, insist on the necessary primacy, as the only hope for holding total chaos and disintegration at bay, of autonomous territorial units – that is states – each pursuing its own interests and, to that end, maximising its own power.

They would echo Machiavelli, who wrote in "The Prince" that

"A prince ought to have no other aim or thought, nor select anything else for his study, than war and its rules and discipline; for this is the sole art that belongs to him who rules" (Machiavelli, p.79).

and Hobbes, when he said that:

"in all times, Kings, and persons of sovereign authority, because of their independency, are in continual jealousies, and in the state and posture of gladiators, having their weapons pointing, and their eyes fixed on one another; that is, their forts, garrisons, and guns upon the frontiers of their Kingdom; and continual spies upon their neighbours: which is the posture of war" (Hobbes, Book I, Chapter XIII).

31

As Stanley Hoffmann put it, amplifying, and apparently endorsing, the views of Rousseau: "Peaceful international politics, of which international law is one aspect, is but the continuation of war by other means" (Hoffmann 1965, p.68). Thus it will be the generally powerful states – those that can convincingly promise and threaten - that will decide the outcome of events, and they will decide it according to their rivalries and interests; the question of what is in the interests of mankind simply cannot be taken seriously. Above all, such states will never contemplate establishing a rival actor, capable of speaking and acting for mankind as a whole, that is not under their control. Since such a body might impose limits on one's own freedom of action, it would be equivalent to voluntarily adding to one's potential rivals and enemies.

It would also be inconceivable, on this view, that the course of history could be changed by an individual like Dr. Pardo, or by a small and powerless state like Malta. A three-hour speech by a man with no divisions behind him can at most cause a ripple in the surface of events, which may extend to the debates and resolutions of the General Assembly and the protracted sessions of the Third United Nations Conference on the Law of the Sea, but what emerges will be determined by whoever can credibly threaten most and promise most. If the Sea-bed Authority is created, it will either be the instrument of one or more powerful states or totally insignificant.

Much evidence to support this can be found in the experience of the League of Nations and the United Nations. The League was never an actor in "high politics". 'The United Nations might have appeared to be in the Korean War, but it was the United States' decision to aid South Korea that provided the coercive power of the so-called United Nations Forces and an American, responsible (more or less) to the American President, and certainly not to the United Nations, that commanded those UN Forces; the most striking episode pointing to the contrary occurred in what is now Zaire, but was in 1960, the Congo, the emergence of a United Nations force of some 20,000 men, at most, that was, through its Secretary-General, capable of acting independently; but it ignited such a constitutional conflagration that the United Nations itself almost perished in the flames, and survived only on the strict, and warily maintained, understanding that nothing like this should ever happen again.

There is another, mellower, view of international relations, which is hardly more hospitable than the "realists" to the "common heritage" idea, whereby states, by recognising each other as sovereign equals, constitute a society with mutual rights and obligations, some, but by no means all, of which are embodied in international law. Hedley Bull has called this the Grotian conception of international society. Grotius himself put it in these words:

> "just as the laws of each state have in view the advantage of that state, so by mutual consent it has become possible that certain laws should originate as between all states, or a great many states; and it is apparent that the laws thus originating had in view the advantage, not of particular states, but of the great society of states. And this is what is called the law of nations whenever we distinguish that term from the law of nature... For since ... the national who in his own country obeys its laws is not foolish, even though, out of regard for that law, he may be obliged to forgo certain things advantageous for himself, so that nation is not foolish which does not press its own advantage to the point of disregarding the laws common to nations" (Forsyth, Keens-Soper and Savigear 1970, p.48).

Thus states might, in general, "behave themselves" — conform to some common standards in their dealings with one another, not just because of what would happen to them if they did not, but also out of a certain self-respect, a consequence of what, as sovereign states, they see themselves to be, that is entities created to reflect honourable and respectable conceptions of what is the common good for a human group.

If, then, what we have now is a world-wide extension of the society of states of Grotius' day, a club whose members are the member states of the United Nations, with one or two outsiders, each caught in a web of mutual expectations, and each normally anxious, except in revolutionary times, both to exact its due from, and to conduct itself so as to remain in good standing with, the rest, then states might well acknowledge the necessity, or the desirability, of jointly devising rules specifying limits of coastal states' jurisdiction over the sea-bed, or, in what remained international, the mode of allocation of mine-sites, and the precautions miners must follow so as not to interfere with other users of the sea; but they would hardly expect such miners (states or companies) to share the benefits of such exploitation with the world as a whole, or to avoid damaging the economies of land-based producers of the same minerals; just as the bonds of membership of a society of states have not yet afforded one producer protection (outside his national market) from competition from another state, or from synthetic substitutes, or required a resource-rich state to share its prosperity with the world as a whole. Still less would such a view entail that states should, or even could, create an agent of "mankind" endowed with the right to exploit the sea-bed directly or control its exploitation by others. In this perspective, the area of the sea-bed beyond national jurisdiction could not meaningfully be called the common heritage of mankind;   but it _could_ be called the common heritage of _states_.

The moral side to this argument has been put with some severity by Hedley Bull. "... in the present condition of world politics ... ideas of cosmopolitan or world justice play very little part at all. The world society or community whose common good they purport to define does not exist ... For guidance as to what the interests of the world as a whole might be ... we are forced to look to the views of sovereign states and of the international organisations they dominate" (Bull 1977, pp.85-86).(1)

The Grotian view of international relations might not follow the realist in ridiculing the idea that new rules (of the first type) might emanate from the proposals of a small and weak state, such as Malta, or from a man of vision, like Dr. Pardo, who happened to be its representative; but their validity would be seen to depend on their being accepted by the overwhelming mass of states including those primarily involved in the activities concerned.

Marxism is another, and more recent, tradition in thinking about the world politics which, while challenging the underlying assumption of the two earlier schools, is, in its implications, equally inimical to the notion that UNCLOS III could make a reality of "the common heritage of mankind". For Marxists, classes, not states, are the fundamental units of world politics. The state is a bourgeois institution, and cannot claim to speak for its people. The proletariat shall inherit the earth, and by means of a violent overthrow of such bourgeois governments.

Marxists originally expected that the success of the revolution in any one country would be a prelude to a struggle to the death between the international proletariat and its exploiters. Among bourgeois states, war reflects, not conflicting national interests, but the ceaseless clash of competing national bourgeoisies; peace on the other hand, represents only temporary unity against a class enemy. (2) Thus Lenin denounced the League of Nations as the "Holy Alliance of the Bourgeoisie", and Chicherin, in a Note to Woodrow Wilson, declared that Soviet Russia would take part (uninvited!) in negotiations about the establishment of such a body, only on certain conditions. One of these conditions was "the expropriation of the capitalists of all countries." For capitalist states to sit down, still unregenerated, in the same international body as the champions of the people, was then unthinkable (though it was soon to happen.) It would therefore seem to be even more unthinkable to Marxists to establish, jointly with capitalist and even feudal governments, a 'regime and machinery' to manage resources declared to be "the common heritage of mankind."

Russia, under Lenin and Stalin, was soon to recognise that the transition from capitalism to socialism would be protracted, and that during that time, the two worlds would have to coexist. It collaborated in many of the activities associated with the League of Nations in the 1920s before joining that body in 1934; and even after being expelled from it, in 1939, it was a founder member of the United Nations and joined many, though by no means all, of its specialised agencies. Stalin justified this participation in the affairs of the bourgeois world by the slogan "socialism in one country"; Khrushchev, by that of "peaceful coexistence". Such cooperation could be seen as tactical, necessitated by the needs of a socialist state in a largely capitalist world; it stopped short of creating a composite actor within which it would only constitute one element. More recently (but dating at least from Khrushchev) Soviet theorists have depicted a threefold division of the world into socio-economic systems, or 'camps': capitalist, socialist and third world, each with its group of "progressive forces": 'the international working class' in the capitalist world; the national liberation movement in the Third World; and the socialist camp itself (Kubalkova and Cruickshank 1977, p.296).

Since existing Third World governments cannot, in most cases, be identified with 'national liberation movements', and Western governments are (in Marxist eyes) even further from representing the international working class, this conception of world politics hardly makes it easier to imagine fashioning an agent for "mankind" out of existing states as now constituted.

Not all Marxists, of course, acknowledged the Soviet Union (as it became) as their model, especially after Mao's success in China, and the Sino-Soviet split, had so manifestly produced an alternative. Some Marxists might deny that either regime, or indeed any so far seen, was genuinely Marxist; if they did that enhanced rather than diminished the absurdity of regarding the states represented at the United Nations and its Law of the Sea Conference as being entitled to set up a body capable of acting on mankind's behalf in respect of his heritage.

There was, however, one thing, at least, about the exploitation of the sea-bed, on which Marxist theorists would tend to agree: that the driving force behind it would be monopoly capitalism, searching for new profits and new supplies of raw materials. The governments of capitalist states would be mere agents of the capitalists themselves; for them, the notion of the common heritage of mankind would be a mere blind behind which the giant consortia would be given the green light to make the sea-bed theirs.

Logically, to the Marxist, the eventual remedy for this unhappy situation would lie in revolution in the capitalist states themselves - indeed, in world-wide revolution. Meanwhile, two tactics might use the concept of the common heritage to promise some mitigation of the domination of sea-bed exploitation by monopoly capitalism: first, the demand that all states, whatever their social systems, had equal rights to sites, thus ensuring that at least some part of the international area benefited the "socialist camp" (assuming there is one!); and secondly, appealing to the developing countries, now constituting a majority in the United Nations and thus in the Conference, to insist on establishing a body, which they could by their numerical strength dominate, which might thwart and paralyse the monopolies. The fact that developing states, whatever their social or economic system, are acutely conscious of the differences that divide them from the capitalist West, makes the unity necessary for such a tactic feasible. Of these two tactics, the USSR adopted the first and China the second.

Nevertheless, each of these tactics poses theoretical problems for a Marxist. To ration each state to a certain number of sites is to emphasise the primacy of states in international politics, which is totally at variance with Marxist theory; but given the existence of at least one Marxist state (if that is not a contradiction in itself) and the necessity for coexistence with capitalism, it make sense as a defensive move. To establish a powerful international authority controlled by developing states, most of which are still dominated by their economic relations with the capitalist world, and can hardly be called "socialist", is to create an instrument which could be used against socialism (as the United Nations was in Korea) and in any case could, as we have seen, have no claim, that a Marxist could accept, to represent 'mankind' as a whole. The objections thus raised, by realists, Marxists and others, to the principle of the "common heritage" might be countered in one of two ways. The first would be to argue that, while there is, as yet, no body that could legitimately be said to represent "mankind", it would be possible, through UNCLOS III, to create one. A sea-bed authority could be the instrument, or one of the instruments, of system transformation. Evan Luard, later to become Minister in charge of British policy at the conference, put it thus: "The newly discovered resources of the sea-bed provide, for the first time in man's history, the opportunity to put into practice a form of world socialism. Here for the first time, the common ownership of common resources, and their use in the interest of all, may be used to bring about redistribution, however inadequate, not from rich individuals to poor within single states, but from rich individuals to poor within the world as a whole" (Luard 1974, p.296).

In other words, by establishing an agent of mankind as a whole, these negotiations might generate a "universal actor", independent of states, though constituted by them, which would "govern" a part of human activity on a worldwide basis. Through it, to an extent not found in the United Nations itself or its specialised agencies, "the whole" might for certain purposes come to prevail over the parts. This prospect would have an obvious appeal for advocates of world government, who seek to create a central locus of power with the capacity to legislate, to adjudicate, and to execute laws and judicial decisions; usually without proposing to abolish existing states, but rather permitting them, like the fifty American states that make up the USA, to retain all powers except those specifically transferred to the centre; thus envisaging, in practice, a world federal government.

World federalists would not, of course, see a sea-bed authority, or even a whole complex of 'ocean space' institutions, such as Dr. Pardo was subsequently to favour, (3) as constituting a world government; only when all states had surrendered to global institutions the right and capacity to make war could that be properly said to exist; but they could see it, if appropriately constituted and authorised, as filling a power vacuum which for technical reasons states had hitherto permitted to exist, and thereby acquiring some of the power and legitimacy that a world government would need. A body established by all the states of the world to govern ocean mining on behalf of mankind would be a world government in embryo, provided that it was given the right, within its field, to uphold global interests, to enforce them (perhaps by a form of sea police) and to require states and other participants such as mining consortia to comply with a global legal regime. Federalists, then, see no contradiction in states declaring some of the earth's resources the "common heritage of mankind"; it is rather an opportunity to create an agent for mankind and give it the powers it needs to act on mankind's behalf. If this requires a conscious decision by states to effect a significant transformation in the international political system, they must be encouraged to take that decision by being persuaded that it is necessary and appropriate to the contemporary world situation.

Some powerful theoretical arguments can be deployed against this straightforward "world federalist" view. To create what they saw as a "world government in embryo" would go against the persistent disposition of most states to retain, in practice even if not in theory, the maximum freedom of action. Rare examples of conscious surrender of sovereignty - such as that of the thirteen US colonies, the merger of Egypt, Syria and later Yemen in the United Arab Republic, or the Ghana-Guinea Union - have often been followed by second thoughts, taking the form of a break-up of the union, or at least the attempt at secession. Such transfer of powers has been particularly unlikely among the most powerful states in any given milieu. They would be in less need of unity as a counteracting weight in their transactions. Add to this general point, about the inclinations of states, a contemplation of the beliefs underlying the major powers of to-day, and the probability dwindles to the microscopic. We have seen why Marxist powers might be expected to resist this concept, and it hardly needs emphasising that the Soviet Union is not likely to abandon its freedom of action in favour of an international body which it is not able to control, even one which did, as Luard thought it might,"provide.. the opportunity to put into practice a form of world socialism"; a prospect the American Senate, for instance, is hardly likely to enthuse over, and which the electoral victories of Margaret Thatcher in 1979 and 1983 and Ronald Reagan in 1980 make even less palatable to the administrations of those two countries. There are however others, besides the world federalists, who allow for the possibility of global integration, notably the functionalists. Functionalism has many variants, going back not merely to the aftermath of the Second World War (with David Mitrany's 'A Working Peace System') but to the First, and even earlier (with Paul Reinsch's 'Public International Unions', Leonard Woolf's 'International Government' and Arthur Salter's 'Allied Shipping Control'). The core of the functionalist analysis is its diagnosis of why the loyalties of individuals, and particularly of 'significant elites', are primarily given to territorial units - sovereign states pursuing national interests. The explanation it gives is that loyalties are given to institutions that best appear to meet human needs, notably economic and social needs. A state will therefore command the loyalties of its citizens so long as no one can satisfy their needs better; it is a self-perpetuating coalition of interests kept domestic by being channelled through foreign offices. Where, however, specific sector interests can be organised internationally, they can often create international interests, embodied in international organisations, which, through understanding, expertise, and management, may satisfy the sectoral interests within each country better than its own government can. If this happens, they will acquire political importance. The monolithic power and authority of the state would be eroded, and sectoral interests would vie with, and perhaps even dominate, national interests.

Functionalists will therefore look at proposals for establishment of a Sea-bed Authority rather differently from federalists. The federalist would ask: can this body validly claim to represent 'mankind' and has it been given the powers it needs to act on mankind's behalf? The functionalist would ask: is this body so constituted as to facilitate the growth of an international interest in this field? Does it, for instance, have adequate (and direct) representation of the main interest groups such as ocean miners, consumers, land-based producers, and environmentalists? Has it got the expertise to work out, among these often competing interests, compromises so efficient and so balanced that none will be tempted to reject them in favour of unilateral national action?

Both functionalists and federalists may therefore see a sea-bed authority as a possible agent of the transformation of world politics; but while the federalist would emphasise its potential uniqueness, as the foundation stone of a new global order, the functionalist would see it as one among many international bodies forging global interests in a variety of fields. Federalists emphasise the surrender of sovereignty, the solemn renunciation by states of their freedom to act within a specified domain; functionalists speak of the 'erosion' of sovereignty as citizens and interest groups increasingly look to global rather than national bodies to meet their needs.

Yet precisely because this conception of the "common heritage of mankind" also sees it as leading, ultimately, to the transformation of world politics, it is also open to the objections raised earlier. States, as such, are not merely propelled by the aggregation of the many diverse interests of the societies they reflect; they also have interests in perpetuating themselves, and those who represent them may not inertly permit their own institutional machinery, the governments of the world and their foreign offices and diplomatic networks, to be superseded or by-passed. It would be more plausible to see international institutions, and a sea-bed authority in particular, in gradualist rather than functionalist terms. In other words, such an authority would be capable of becoming an important actor in world politics, but only if states (and not just certain relevant interest groups within them) retain their confidence in it. Sea-bed mining in "the common heritage of mankind" cannot be either wholly internationalised (as the federalists might hope) or wholly depoliticised (as the functionalists would be tempted to assume). Gradualism (which I have called "political functionalism") (Ogley 1969, p.617) argues that, to be a significant international actor, an international organisation has to be both technically competent and politically balanced, and that decisions to endow it with important powers must be "reversible". Applied to the constitutional crisis which the United Nations underwent as a result of the Congo episode, this offers a subtler perspective than the "realist" explanation given earlier, without disagreeing fundamentally about why it nearly brought the United Nations to the point of disintegration. The divisive effects of the UN force in the Congo (ONUC) arose both because it appeared to affect the outcome of the internal conflict between Kasavubu and Lumumba in a way that favoured one outside interest (the West) at the expense of another (the USSR), and because, when the international consensus which launched it broke down in consequence, there was no mechanism whereby the dissatisfied minority could terminate it. Subsequently UN forces in Cyprus (1964) and the Middle East (1973) have avoided divisive consequences of this kind by being authorised for a limited period, but renewable. Thus states (especially states with vetoes) that vote to set up such forces can later change their minds. International cooperation governed by this principle of "reversibility" permits those who might one day "attempt from love's sickness to fly" to do so; in contrast with traditional functionalism, political functionalism, or 'gradualism', relies not on by-passing foreign offices but on developing habits of cooperation among them. Accordingly, there is more chance of realising the "common heritage of mankind" through arrangements that are tentative in character, and do not, in themselves, imply an irreversible commitment to a given set of norms institutions or procedures.

Not dissimilar conclusions could be reached through two further approaches that have sought to demonstrate that there is more to international relations than can be reduced to the interactions of states (or indeed of classes): the cybernetic, as expounded and applied to international relations by Karl Deutsch in "The Nerves of Government"; and that of "complex interdependence" developed by Robert Keohane and Joseph Nye. The cybernetic approach emphasises the need for communication and feedback in decision-patterns. Clashes of interest are not inevitable among sovereign states; they arise through the unthinking adoption by one of policies that frustrate others' needs, and inability to change policies easily once made. Concepts of "lead" (how far an actor can see ahead), "lag" (how long it takes an actor to respond to new information appropriately and implement its response) and "load" (how many demands compete for its attention at any one time) are prominent in the assessment of its capacity for self-steering. On this view, organisations can help states to be autonomous, and more responsive to each other, and this combination is not contrary to the nature of states (as realists would suppose). A sea-bed authority would be important in proportion not to its powers (as the federalists would judge it) but to its capacity to see ahead and help member states and companies to see ahead too. It is a way of organising those involved in sea-bed mining or affected by it so that they can give adequate attention to the problems it is like to generate and satisfy their collective ambitions as fully as possible. The emphasis is on learning, and autonomy. Though close to the notion of a society of states, capable of recognising and abiding by rules necessary for this coexistence, the cybernetic approach is distinctive because of the emphasis it puts on communication patterns. Cybernetically, the point of calling some domain a "common heritage" would be to widen the range of entities, political, commercial, environmental or any other, whose views, attitudes, and welfare would be given some legitimacy by those involved in mining it. The implication is that, on the one hand, miners would be internationally inspected, and submit, in international fora, reports on, and where necessary explanations of, what they were doing; on the other, the world would survey the industry, land-based and sea-based, and suggest the likely consequences, on a continually-updated global basis, of alternative policies.

Robert Keohane and Joseph Nye's notion of complex interdependence, developed in their book, Power and Interdependence, which in parts derives from a consideration of the early stages of UNCLOS III, lays great stress on the notion of an "issue area".

This might be the law of the sea as a whole, or, more narrowly, sea-bed mining. Each issue area has its own regime, whose development is determined by four factors: the general power of states (and the importance of the issue area in question to the most generally powerful states); the specific power of states within the issue area in question (such as the ability to enforce or to resist the application of jurisdiction within a maritime zone, or to mine the sea-bed unilaterally if need be); considerations of economic efficiency in the broadest sense, that is the net contribution of a regime to human welfare in situations of changing knowledge and changing technical options; and "organisationally dependent capabilities" - that is the set of networks, norms and institutions within which the issue area is initially handled. Thus, argue Keohane and Nye, the existence of the United Nations in the form which it had in 1967 and as it developed thereafter framed the law of the sea issue area in a context of "egalitarian organisational procedures" and confrontation between rich and poor.

This context, and the question of whether Keohane and Nye have correctly characterised it, will be examined later. (4) The point is that, within it, the notion of "the common heritage of mankind", whether or not it can be given any intrinsic meaning, can be seen as a factor precluding certain outcomes and promoting others. The institutional dynamic carries the negotiations in a certain direction, which even a powerful state can resist only at some cost. In the absence of such resistance, the inertia of the negotiating process will tend to produce a regime which gives some content to the "common heritage of mankind", as understood in its context. This regime, and any institutions associated with it, will survive so long as they are not seriously challenged by a dissatisfied power. Keohane and Nye do not say that complex interdependence applies in all issue areas. They see some as more directly governed by naked power. But, where it does apply, "organisationally - dependent capabilities" will shape both what may be legitimised (i.e. what rules are compatible with the prevalent doctrine) and how (i.e. who participates in regime change?).

The "common heritage of mankind" as applied to the sea-bed is thus an idea on the borderline between dream and reality. It is novel, and would be ruled out of court by realists or, as conceivably capable of emerging from the world's existing political and economic structure, by Marxists. Its very resonance conjures a vision akin to that of the world federalists, which may both have aroused expectations and nursed suspicions; yet other less implausible modes of putting it into effect are feasible. It does not have to be seen as generating a powerful international authority, only an effective one. It is indeed a task and an opportunity that cries out for international collaboration and institutional inventiveness; but such collaboration might have the most lasting political effects if, rather than insisting on devising something that from the start would be as perfect a realisation of the concept as possible, it concentrated rather on building flexibility, review, and adjustment so pervasively into the initial embodiment of the idea, that none could see themselves as being asked to commit themselves to what they might fear would prove an irretrievable error.

NOTES

(1) Presumably, if an international organisation, though set up by states, was not dominated by them, it would forfeit its claim.

(2) "in the international sphere it (the bourgeoisie) is ceaselessly in conflict with the bourgeoisie of foreign countries ... as a result of 'foreign policies in pursuit of criminal designs, playing upon national prejudices and squandering in piratical wars the people's blood and treasure.'" (Kubalkova and Cruickshank 1977, p.292, quoting Marx, "The Civil War in France", Selected Works in One Volume, London, 1968, p.260).

(3) e.g. in his speech, of March 23 1971, still on Malta's behalf, to the Sea-bed Committee, (Pardo 1975, p.189).

(4) Chapter 5.

4 The Participants

Awareness of the new possibilities of sea-bed mining, beyond the undisputedly coastal domain of the continental shelf as geographically defined, prompted two kinds of response: a search, initiated as we have seen by Arvid Pardo, for a regime for the global management of this new kind of activity; and plans, by those who saw themselves as capable of exploiting these possibilities (private companies, states, and combinations of the two), to develop the means to do so.

The latter are important participants in the process of creating a regime, for two reasons: first, if the regime is too onerous, or too uncertain, they may abandon their plans, or drastically reduce them; and secondly, they may claim, with varying degrees of credibility, that they can mine in disregard of it. Legally this claim is associated with the proposition that mining the sea-bed, beyond the limits of national jurisdiction, is a "freedom of the seas"; in practical terms, sea-bed miners have sought title to mine a specific area of the sea-bed - a "site" - for a specific number of years. Because, as with all mining, initial investment is heavy and there is a a long gap before profits are made, each sea-bed miner needs to know that no other mining ventures will impinge on any part of the area claimed, and that no states will exercise jurisdiction, to their detriment, or interfere, deliberately or accidentally, with their activities, including associated activities such as transport and marketing. If their title is likely to be put seriously at hazard in one of these ways, they may not mine, and indeed if they are private firms may be unable to raise the necessary capital to do so. This may lead them to seek title from states, and thus the passage of national legislation. The question then arises as to whether a protecting state, or a state envisaging mining on its own behalf, is prepared to incur the costs and risks of defending a mining operation against these hazards. This in turn may depend on which other states are prepared to offer similar protection outside a global regime and whether among such states mutual recognition of claims can be negotiated. The uncertainty can even be taken a stage further, since the willingness of states to contemplate a less than global regime is itself dependent on how widely-supported that global regime is and what its supporters are prepared to do to impede ocean mining outside its jurisdiction. Moreover, this causal chain can be reversed. The extent of support for a global regime and the leverage it can wield against dissenters depends on how many, how powerful, and how determined the dissenters are. Thus it makes a difference to the chances of a global mining regime whether prospective miners, even if they are not themselves states, are prepared to accept it.

In the field of nodule mining, which has been the central preoccupation of those negotiating a sea-bed regime, Buzan (1976, pp.151-153) gives a good summary of commercial developments up to 1974, listing four US companies (Deepsea Ventures; Kennecott; Ocean Resources; and Summa Corporation, the latter including Lockheed and Global Marine); two Japanese (Sumitomo and Deep Ocean Mining Association); one West German joint venture composed of three companies with financial help from their government; two French companies in partnership (CNEXO and Societe le Nickel); and two Canadian companies (INCO and Noranda Mines). A consortium of 32 companies (seven American, including Ocean Resources; one German; the two French companies mentioned; one Australian; six Canadian, including INCO; and the rest Japanese, under the umbrella of the Sumitomo Group), had promoted tests of the Continuous Line Bucket System invented by Commander Masudo of Japan in 1966. At this stage there were no British, Italian, Belgian or Dutch companies participating, and no sign of Soviet development of a nodule mining capacity.

By 1980 six consortia had formed, in addition to the Continuous Line Bucket Syndicate in which many of the constituents of the consortia had already participated. Of the consortia, one (AFERNOD) is wholly French, consisting of the two original companies and three governmental agencies (Le Commissariat à l'énergie - CEA; Chantiers de France - Dunkerque; and Bureau de recherches géologiques et minières - BRGM); one wholly Japanese (Deep Ocean Minerals Association, embracing thirty-five companies, some of which also participate in other consortia); and four are complicated international mixtures. Two of the earliest companies, Kennecott which began research in the field in 1963 and Deepsea Ventures which started in 1968, have been transformed. British firms (Rio Tinto-Zinc, Consolidated Gold Fields, and more recently British Petroleum) became partners of Kennecott Copper Corporation in the Kennecott Group and subsequently one of them, B.P., secured a controlling interest in the consortium by indirectly taking over Kennecott Copper itself (via Standard Oil of Ohio, in which BP has a majority share). The consortium also includes a Canadian and a Japanese company (Noranda Mines and the Mitsubishi Corporation).

Deepsea Ventures was originally a subsidiary of Tenneco, and a successor to Newport News Shipbuilding and Drydock Company, which has been in the business since 1962. In 1974 it went into partnership with a group of five Japanese companies (collectively the Japanese Manganese Nodule Development Company (JAMCO)) and later in the same year these were joined by US Steel and Union Minière of Belgium; the four became Ocean Mining Associates. Later Deepsea Ventures disappeared as Tenneco (and JAMCO) pulled out. The two others were joined by Sun Co., also American, and, according to Dr. Post (Post 1983, p.216), an Italian firm Sanim, of the ENI Group. US Steel has a strong interest in manganese mines in Gabon and South Africa.

INCO, a Canadian firm holding a near-monopoly, world wide, in nickel mining, was in the field even earlier, in 1959. In 1975, it joined with Deep Ocean Mining Company (DOMCO) of Japan, AMR of West Germany (consisting of Metallgesellschaft, Preussag, and a government-owned firm, Salzgitter) and shortly afterwards SEDCO (USA), to form Ocean Management Incorporated.

The third of the four American companies mentioned by Buzan, Ocean Resources, was founded by John Mero in his attempt to prove that nodule mining could be undertaken with modest capital expenditure, using a CLB system. By 1980 it seems to have retired from the field and is not among the constituents of the six major consortia listed by the UN.

Summa Corporation, the fourth American firm on Buzan's list, was associated with Howard Hughes, Lockheed and Global Marine. It took the limelight in 1975 with the use of its nodule mining ship in an unsuccessful attempt to raise a Soviet submarine which had sunk in the Pacific. In 1977, Lockheed's Ocean Minerals Inc., which included two Dutch firms, Billiton and Royal Boskalis, joined Amoco Ocean Minerals Co.(a subsidiary of Standard Oil of Indiana) to form the Ocean Minerals Company. Billiton is a subsidiary of the Royal Dutch/Shell Group.

It will be seen that many of the components of the ocean mining consortia have interests in land-based mining of the same metals. They differ, however, in their attitude to the creation of an international authority. Ocean Mining Associates, particularly in its Deepsea Ventures days, was the least hospitable to the idea, and in 1975, Deepsea Ventures filed with all governments a claim to a 60,000 sq.km. site in the Pacific; and in the same year its President, John Flipse, threatened his country that unless there was

"a good law on the books of the United States, with enough protection and enough incentives to offset the taxes and the restrictions found in the United States . . . . we will go foreign"

(Christy, Clingan, Gamble, Knight, and Miles 1975, p.331).

He and the company's lawyer, Northcutt Ely, vied with each other in the contempt with which they spoke of the United Nations and UNCLOS III.

Kennecott, by contrast, led, in its ocean mining activities, by Marne Dubs, has tended to be more restrained and judicious. When Carter took office in 1977, Kennecott acquired Leigh Ratiner, who had lost his job on the US Delegation, as counsel; and both men were extremely critical of Richardson's handling of negotiations; in a letter to all Congressmen on May 26 1978, for instance, Ratiner enclosed a speech of Dubs which, while affirming that

"a good treaty on the Law of the Sea is better than no treaty at all",

insisted that "fundamental changes" were needed in the US Delegation's instructions if a treaty acceptable to the USA, and particularly its Senate, were to result. Yet in 1982, Dubs, like Ratiner, dissociated himself from the US rejection of the convention, and held that, given the fulfilment of certain conditions, industry could operate under it (Citizens for Ocean Law 1982).

Thus in the years 1973 to 1983 British, Belgian, Dutch and Italian firms have appeared on the ocean mining scene; a British firm, BP, has acquired a dominant interest in one consortium; in only one group (OMCO) do US companies hold a majority interest, and in another (OMA) two American companies each hold a quarter of the shares. In the private sector, eight countries: Belgium, Canada, the Federal Republic of Germany, Italy, Japan, The Netherlands, The UK and the USA, have companies involved although, as John Flipse showed, some of these are quite prepared to change their allegiance in pursuit of advantageous terms. In addition four states claimed pioneer status by virtue of purely national enterprises: France, India, Japan and the USSR. India was a surprising late addition to the list of ocean mining states, the only "developing country" among them so far, and hitherto, one that had advocated strict control and heavy tax burdens on such miners. The French group is a mixture of private and state elements, as is the Japanese, who of course, as we have seen, are also represented among the private consortia.

Besides prospective miners, private or state, and in the private case, the states that represent them, the main participants in the process of regime-making are states in their other guises: importers and exporters, current and potential, of the metals likely to be produced; coastal states, interested in expanding the area of sea-bed under their jurisdiction, at the expense of the "common heritage", as opposed to land-locked and "shelf-locked" states, who have nothing to gain from such extensions and insofar as they stand to benefit from revenues derived from the "common heritage", everything to gain by making the latter as large as possible; and finally states likely to be financial contributors insofar as the Authority needs money, and those likely to be recipients of revenues and technology.

It can be seen that, within each of these pairs of opposed interests, bargaining power is not symmetrical. Ocean miners can threaten to 'go it alone', and consumers to support them, if land-based producers ask for too much protection. Existing producers are vulnerable to such competition; even if one or more states enjoy a near monopoly in a metal, they can exploit that power only so long as ocean mining does not take place. Coastal states can assert jurisdiction by force over further and further reaches of continental shelf. Land-locked and shelf-locked states may deplore this trend but, as such, have no counter threat. There is no common power to defend the common heritage. Finally, and even more obviously, potential contributors of funds and other assets to the Authority can threaten not to join and thus to avoid having to make such contributions; potential recipients have to accept in the end the best deal they can get.

In the setting of UNCLOS III, however, all states, whether or not they have any external bargaining power, have a say and a vote. A regime can gain the legitimacy that flows from being a UN Convention only if it secures the appropriate majority of the votes cast - in this case two-thirds, to include a majority of the states participating. Thus certain categories of states, because they are sufficiently numerous to make a difference to the chances of achieving such a majority, can also exert bargaining power. Here the Group of 77, consisting of the developing countries of Africa, Asia, Latin America (and of the Caribbean and the Pacific), is in a particularly strong position. In pure voting terms, it can, on its own, if united, dictate the terms of a convention: from the beginning of UNCLOS III its members have comprised more than two-thirds of the participating states; and as most subsequent additions to the conference have been ex-colonies this proportion has increased.

Objectively, one might expect the interest of developing countries to lie in securing the maximum yield to the Authority from the exploitation of the "common heritage", since they are likely to be revenue recipients rather than revenue payers. In fact they have shown even greater interest in two other issues: participation, through the Authority, in the actual process of ocean mining; and the transfer of technology both to developing countries and to the Authority itself. But if private and state mining is to provide revenues and technology (and sites), there has to be such mining. The interests of the Group of 77 as a whole, therefore, do not lie in strangling ocean mining at birth (though that might benefit the land-based producers among them) but in setting the highest price for it that the industry will bear. There is, naturally, some disagreement as to what that price is.

The issues between coastal states and "the land-locked and geographically-disadvantaged" cut across the developing country-developed country dimension. There were in all 29 land-locked states at UNCLOS III, and at least 22 shelf-locked. The term "geographically-disadvantaged" included the latter, and also others restricted by geography from extending their jurisdiction, such as zone-locked states who could gain little if anything from a 200 mile exclusive economic zone. Together they were able to amount to something like a blocking third; and thus, theoretically, to threaten to deny legitimacy to a convention that did not satisfy them. This was not, however, a very formidable threat. For it to be credible all such states would have had to agree on what improvements they wanted, and to have seemed, at least, to be prepared to vote for a convention only if these were made; but since geographically-disadvantaged states often had divergent other interests it was always very unlikely that they would all finally judge the Convention solely on these grounds.

The question of whether land-based producers should be protected against sea-bed competition is another issue that complicates the developed-developing country division, in that some major exporters of nodule minerals, notably Australia and Canada, are rich developed countries, and that for only a handful of developing countries are the nodule metals a significant element in exports. To these could be added potential producers - countries that might expect one day to be enriched by the exploitation of hitherto untapped resources within their territories in the absence of ocean mining; and exporters of substitutes for the sea-bed metals, the markets for which could also be affected by sea-bed mining.

On such issues, the Group of 77's position is far from being simply that of championing the common heritage. Other factors are at work. Coastal states, for instance, interested in reserving for themselves the fish stocks near to their coast, and regulating the activities of maritime powers in these waters, may tend to see these causes as strengthened by a rule drawing the boundary of their sea-bed jurisdiction far from their coast, and thus reducing the size of the common heritage: and developing countries not involved in that issue, as such, may tend to support them since, on the whole, distant-water fishing, navigation, and military activity tend to be the preserve of developed states and therefore easily acquire an "imperialist" tinge. Hence the widespread developing-country support for the exclusive economic zone discussed in more detail in chapter 6. Again, seen as commodities, the metals of the sea-bed set precedents for other commodity negotiations. In these negotiations,the exporting countries are usually developing countries and the major importing countries developed countries, and certainly, on balance, such commodities flow from developing to developed countries rather than vice versa. The maintenance of stable and remunerative (i.e. higher in the long run than might result from market forces) prices for such commodities is part of the programme of the NIEO and, since Nairobi in 1976, of UNCTAD. Thus developing countries which export quite unrelated commodities may derive benefits from setting a precedent, with respect to sea-bed minerals, of firm control over production and prices. Less hard-headedly, the developed world is seen as "exploiting" the developing world by encouraging the latter to specialise in primary products, another dimension of what is seen as their imperialism.

Interests in the possible precedents these negotiations could set are endless. For the Group of 77, a reputation for firmness is valuable for other negotiations even if it fails to produce results in this one. The same could be said, of course, of their opposite numbers in the West. Similarly, both sides are also concerned about the powers of international organisation, and the structure of decision-making in them, that a given outcome might suggest as a norm in other cases. To concede a form of explicitly-weighted voting, here, was seen by developing countries as a clear pointer to future international bodies (though in fact a form of weighted voting was agreed in 1980 for UNCTAD's scaled-down Common Fund). A strict procedure for the settlement of disputes about sea-bed mining was seen by some developed states, and particularly the USA, as even more valuable, insofar as it might also lead to the more general acceptance of recourse to arbitration, judicial settlement, or other mechanisms, in law of the sea issues generally.

The process of considering the "common heritage" question since 1967 has divided states into four categories, roughly equal in numbers. The members of the Sea-bed Committee before 1970 (42 from 1968-70); the new members added to it in 1970 and 1971 (48); the 45 other members of the UN who had at least, before UNCLOS III was convened, been able to participate in debates in the General Assembly's Plenary and First Committee; and 35 more states that entered the fray only during UNCLOS III itself, either because, like Switzerland, they were not UN members, or because they gained their independence after 1973. Table 4.1 indicates the distribution of these four categories of state by regional group and where available by other categories discussed above. It will be seen that initially the developed states of the West and East were well-represented, making up three-sevenths of the Committee, but in the major enlargement of 1970-71 (all but five being added in 1970) developing countries took more than three-quarters of the new places, bringing them up to just over two-thirds of the

Committee's overall membership. Land-based producers were well-represented throughout but land-locked states, and least-developed countries, were initially grossly under-represented, and in spite of quadrupling their membership of the Committee, in both cases, in 1970-71, remained substantially so. Thus the Sea-bed Committee never became the General Assembly in microcosm.

A full count of the actors and interests involved in this intricate process of negotiation would have to go beyond states to examine the diverse and often conflicting concerns within any given state. Nowhere is this more visible, or more significant, than in the USA. Ann Hollick has described the bureaucratic struggles that preceded the adoption of the US Draft Treaty as the basis of American Oceans Policy in 1970. Then

> "ocean bureaucrats enjoyed substantial freedom from high level supervision when consensus could be reached among the agencies or once the basic decisions were made . . . until 1973 a rather small group of ocean experts [was] formulating policy isolated from other policy areas as well as from White House scrutiny"
> (Hollick 1974, p.14).

She identified three interests, in addition to the sea-bed miners discussed earlier, who were significantly influential in ocean policy up to 1972: the petroleum industry, the military, and the marine science community. The first two of these were particularly pertinent to America's policy towards the "common heritage". The Defense Department, wanting above all to secure recognised transit rights through straits (including the right of submarines not to surface in transit), innocent passage for warships in the territorial sea, and freedom of navigation beyond twelve miles, was anxious to obtain a comprehensive treaty, and willing to include in it a regime for the "common heritage" that would make it attractive to developing states. Since the department also feared "creeping jurisdiction" on the part of coastal states, it also favoured narrow limits to coastal state jurisdiction over the shelf. In both these respects, therefore, the Defense Department (like the State Department) was for a long time among the staunchest friends of the "common heritage" within the American political system. They were opposed on limits by the oil industry, which did not want any part of America's own potential sea-bed oil to fall into the international realm and, on the regime for the international area, most of the hard minerals industry, that is the companies described earlier, and the Department of the Interior. The fishing industry came in later and proved divided on limits to fisheries jurisdiction, the West Coast favouring narrow limits and the East Coast broad ones. Given this division, US opposition to the emerging clamour for a 200 mile exclusive economic zone evaporated. On the regime itself, the Treasury, in 1973, entered the scene, under George Shultz, and evaluated sea-bed policy from an economic, and nationalist, viewpoint, questioning (in unison with the hard minerals industry) the principle of sharing revenues with the international community, sounding the alarm about the possibility of a Sea-bed Authority's discretionary powers being used to impede efficient use of resources, but opposing the element of subsidy implied in the mining industry's demand that it be guaranteed against financial loss (Hollick 1974, p.61). These elements tended to be reflected in Congress as well as in bureaucratic in-fighting. Some of them, though nominally attached to the US Delegation to UNCLOS III, were able to use its sessions, according to Keohane and Nye, to further their sectional interests in disregard of US oceans policy through appropriate alliances with delegates from other states. (1)

"Actors within actors" can also be discerned with reference to other states, if not quite so openly, or so weightily, as in the USA. For instance, Barbara Johnson and Mark Zacher, in 1977, brought together a number of analyses of the making of Canadian Law of the Sea Policy which shows complexities similar in kind, if not in degree, to those of its southern neighbour (Johnson and Zacher 1977).

Two further types of actor must be enumerated: international government organisations (IGO's) and non-government organisations. The list of delegations to the Caracas session, in 1974, included observers from two other offshoots of the UN (UNCTAD and UNEP), eight specialised agencies and seven other international organisations, notably the Commonwealth and the European Community. Of these UNCTAD's was perhaps the most prominent role. Its studies of markets for some of the nodule metals were presented to the First Committee by an UNCTAD representative, Mr. Arsenis, and stimulated wide interest and discussion, and though their conclusions could be disputed, their scholarly tone secured for them a widespread respect.

On the whole, though, international organisations had little more than the inherent persuasiveness of any arguments they might offer, and the regard in which their representatives might be held, to further their aims. The same is even more true of non-governmental organisations, of which more than thirty are listed as being represented, at one time or another, at Caracas. Some of these established close relations with a wide range of delegations; others, like the Friends World Committee for Consultation and the Ocean Education Project of which it was part, also put on meetings or receptions (and even on occasion weekend conferences) which attracted a fair number of delegates; and joined in publishing a lively and very well-informed newspaper, NEPTUNE, which was widely read by delegates. Several participated in the general debate at Caracas in

Committee I (and one, on behalf of NGOs generally, in the plenary). On the whole, though, delegates did not behave as if NGO's as such had to be taken into account; and they were excluded from "informal" meetings of the First Committee and its various groups. After Caracas, NGOs did not address the Conference again and many of their representatives attached themselves to national delegations. Alberto Szekely (International Student Movement for the United Nations and the NGO's collective spokesman in plenary) joined the Mexican delegation; Elizabeth Borgese (International Ocean Institute, Malta, which also served as a vehicle for Dr. Pardo's views when his country dropped him from their delegation), the Austrian; and Edgar Gold (International Law Association) the Canadian. One can perhaps infer that the advantages of access to conference meetings and papers outweighed the disadvantages of seeking national partisanship; which still reveals something about status accorded NGO's. The ILA's representation also included the future author of the Deepsea Ventures' unilateral claim to a site, Northcutt Ely.

There remains one actor, in all this, that has not yet been discussed, the United Nations itself. The decisions of the United Nations as to the handling of Dr. Pardo's proposal, the holding of the conference, and its organisation, played an important part in shaping the outcome, and will be dealt with in the next chapter.

NOTES

(1) eg. the oil industry and Dept of the Interior lobbying against US official policy for a 'broad' shelf at the Seabed Committee (Keohane and Nye 1977, p.149).

TABLE 4.1

UNCLOS III: Participation at Each Stage.

	AFRICA	ASIA	EASTERN EUROPE	LATIN AMERICA	WESTERN EUROPE AND OTHER	TOTAL
SEA-BED LL*	0	0	1	0	1	2
COMMITTEE SL	1	3	2	0	1	7
1968-70 OS	10	3	3	7	10	33
TOTAL	11	6	6	7	12	42
Land-Based Producers	1	1	1	4	2	9
Least-Developed Countries	2	0	–	0	–	2
ADDITIONS LL	2	2	1	1	0	6
1970-71 SL	2	4			4	10
OS	11	6	(2)	9	4	32
TOTAL	15	12	3	10	8	48
Land-Based Producers	5	3	1	1	3	13
Least-Developed Countries	4	2	–	0	–	6
OTHER LL	11	3	0	1	1	16
UN SL	0	7	1	0	1	9
MEMBERS OS	4	4	1	7	4	20
TO TOTAL	15	14	2	8	6	45
1973 INC.						
Land-Based Producers	2	0	1	2	1	6
Least-Developed Countries	11	4	–	1	–	16
OUTSIDERS LL	1	–	–	–	4	5
AND COASTAL	9	13**	–	7	1	30
NEWCOMERS TOTAL	10	13**	0	7	5	35
Land-Based Producers	1	1	–	–	–	2
Least-Developed Countries	0	2	–	–	–	2
GRAND TOTAL:	51	45	11	32	31	

```
* LL = land-locked
  SL = shelf-locked
  OS = open shelf
** includes S. Vietnam to 1975
```

53

5 The Negotiating Process

To construct, out of the dense mass of separate and contentious interests just described, a new regime, and a new worldwide institution, that could effectively govern an as yet embryonic industry, was an undertaking of daunting complexity.

It absorbed an enormous amount of diplomatic time. UNCLOS III alone, as we have seen, lasted for 93 weeks, spread over eleven sessions; before that, the Sea-bed Committee (taking together its "ad hoc" and "standing" versions) had had fourteen series of meetings from 1968 to 1973, interspersed with regular consideration of its work in the General Assembly. Such prodigious outlay of time and money naturally prompts the questions: was it, in the light of its results, worthwhile? could it have been done more economically? what lessons can be drawn from it which might benefit future attempts to negotiate global solutions to global problems?

For Ed Miles, writing after only the first substantive session of UNCLOS III at Caracas in 1974, the lesson was already clear.

"The General Assembly should _never_ _again_ convene a conference of the size and complexity of the Third Conference on the Law of the Sea. As a decision mechanism it is absurd and in its size and complexity imposes demands on delegates which in their totality are quite beyond the competence of human beings to manage" (Christy, Clingan, Gamble, Knight and Miles 1975, p.40).

Miles attributed the 'extremely slow pace of the conference' to "a concatenation of at least eight separate factors": 1. The size of the conference. 2. The number of issues to be negotiated. 3. The complexity of these issues. 4. The structure of the Group of 77. 5. Inadequate preparation by sub-committee II. 6. The absence of a deadline for decision. 7. The Rules of Procedure. 8. Certain features, including ideological conflict, of Committee I.

Although some of these features, such as the structure of the Group of 77, are external to the conference, the implication is that a working regime covering at least part of the agenda of UNCLOS III, including the exploitation of the deep sea-bed, could have been arrived at more speedily by other means. Others, by contrast, have seen the UNCLOS III process as an "irremediable cost of doing business on global issues in an increasingly interrelated world system", the only way in which the issue raised by Dr. Pardo could have been treated at all, if it were not to be settled by default in chaos, disorganisation and even violence, and predict that "such global conferences in fact will become the dominant and standard vehicle for international law making" (Eustis 1977, p.256).

One thing is clear. Procedures do affect outcomes. As one commentator puts it, "for a given attempt at group decision-making the ultimate outcome depends to an important extent upon the given procedure employed to aggregate the preferences of individuals into 'preference of the group' ... the UNCLOS process is not a 'neutral midwife'" (Eckert 1978, pp.262-3).

It is therefore illuminating to examine just what this process was, and to speculate as to what different results there might have been if the same collection of states and other actors, with the same assortment and divergencies of interests and resources, had attempted to reach agreement by different procedures. To go straight to the hub of the question, if, in fact, in Richardson's time there was enough common ground between the USA and virtually all the rest of the world to permit a convention to be negotiated which they could both, however reluctantly, accept, then why, in those four years, were these negotiations not consummated?

NEGOTIATIONS IN INTERNATIONAL ORGANISATIONS

The very fact of raising a question in an international organisation, and particularly a global one of near universal membership and general scope, has consequences. It means that such a question is treated differently, in at least three ways, from what it would have been if it had been handled through traditional diplomatic channels. First, it prods all members of the organ in which it is raised – and when it is the General Assembly that means all United Nations members – to say where they stand on the issue. They can, if they wish, say that they do not consider it yet ripe for discussion, but they cannot, except by so persuading the body responsible for determining the organ's agenda, (1) prevent it being discussed by those who want to discuss it, and a refusal to discuss, or mere silence, can have its costs. Traditionally, a state seeking joint action on a question proceeds by first consulting those whose response makes the most difference to whether that joint action is possible or successful. Peripheral states, that is those seen as having no direct interest in, or ability to affect, the question, may be told of it, if at all, only when the basis for action by the principals has been agreed. Courtship demands privacy. Institutional consideration makes public what might otherwise be private; it forces states either to repudiate (sincerely or not), or to tacitly acknowledge, intentions they might otherwise have preferred to conceal. It thus short-circuits the diplomatic process. Secondly, because the United Nations has a fairly substantial secretariat, the act of raising an issue in the UN allows for studies to be commissioned and made available to all members. There had been commissioned Secretariat studies of the mineral resources of the ocean floor before Malta's item had been inscribed on the agenda of the General Assembly, (2) but thereafter their flow was greatly accelerated. (3) Even if the Secretariat does not produce information on its own initiative - and there are occasions on which it does so without having been specifically asked for it (4) - the fact that it _can_ be asked to produce a survey, map, or study makes General Assembly debates more than mere pointless confrontations. Carefully-worded as such products will almost always be, they have their political effects, in part, but by not means entirely, through educating delegates; a process which is reinforced by more informal contacts between delegates and the Secretariat. Once an item is on the Organisation's agenda, it becomes the Secretariat's task to help every delegate to be as well-informed as possible, and, in general it is those who are less-informed on a question that benefit from this help. But in addition to reducing (though by no means eliminating) the gap in information between the most knowledgeable (normally the most developed) and the least, what the Secretariat publishes necessarily selects issues, makes assumptions, highlights possibilities, and thus contributes to what is seen as politically important and what is not. In other words, Secretariat studies create saliency, focus perceptions, and, if only marginally, change attitudes. Finally, international consideration of an issue transforms the language in which it is discussed. The language of the United Nations, and particularly, of the General Assembly, is the language of transcendence. Positions must be related, however loosely, to concepts of world peace, international political and economic justice, or the ending of imperial domination. To this must be added, in the case of the question with which this book is concerned, that of the "common heritage of mankind".

The need to debate an issue in such an elevated tone works two ways. On the one hand, it allows global solutions to be discussed. Without the United Nations, and the pressure its procedures generate, it is doubtful whether the question of setting up a sea-bed authority, to enable the world to manage, in some sense, and benefit from, the exploitation of the resources of at least part of the world's ocean, could ever have been considered. Certainly it is hard to imagine that such deliberations would then have preceded the exploitation itself. At the same time, though, it obfuscates the process of negotiations by inhibiting states from conveying the real reason for their positions, if these cannot easily be reconciled with the transcendent language of the debate. UNCLOS III has thus, like other conferences, been prone to resort to private meetings; some open to all participating countries, yet subtly different in tone from those held in public: others, more exclusive, consisting of anything from two or three delegates in a quiet corner of the Delegates' Lounge to the hundred or so members of the Group of 77. But United Nations diplomacy, even in such private meetings, is different from diplomacy outside the UN, insofar as these meetings are ways of arranging support for, or opposition to, a position about to be considered by the conference; and when so presented, they have either to conform, if only nominally, to what the conference has, explicitly or implicitly, indicated as potentially acceptable, or challenge that, at the risk of letting the conference fail. The powerful cannot simply arrange things coolly to suit themselves.

THE UNITED NATIONS IN 1967

These three characteristics: the ease with which a matter can be raised "prematurely", the ability of the secretariat to educate the less well-informed, and the language of transcendence, are all reasons for the powerful to distrust or at least be wary of global international organisation. To these general determinants of great power attitudes, and organisational influence, must always be added the specific composition of the body in question, and its attitude towards particular states or group of states, at the time in question. The United Nations, in 1967, had 122 members, increasing to 123 with the admission of Southern Yemen in December of the same year. The alignment of these states on Cold War issues was in a period of transition. An "automatic Western majority" was still maintaining the exclusion of the People's Republic of China, and the one-sided consideration of the Korean question, in which South Korea but not North Korea was invited to participate; but it had not been automatic enough to support the US in 1965, in seeking to suspend, by virtue of Article 19, the voting rights of the Soviet Union and others who, because of their refusal to pay their share of the expenses of the Congo and Suez operations, had become more than two years in arrears with their contributions. The fact that the costs of these operations had been retrospectively declared "expenses" of the UN, in the teeth of Soviet opposition, was itself a testimony to the anti-communist tendency of the General Assembly (though the legitimacy of such a decision was to be buttressed by an Advisory Opinion of the International Court of Justice); but the fact that it was not taken to its logical conclusion was a sign that such influence was on the wane.

At the same time, the preponderance of developing countries in the General Assembly, and their organisation into a bloc capable of acting with some cohesion, was by now established. On colonial questions, the Assembly's resolution of 1960, calling for the termination of all forms of colonialism, was a ratification, in form, of the trend of the preceding quinquennium, which had seen the independence and acceptance into the UN of most sizeable European-rule territories of Africa and Asia. (5) The Committee set up under the 1960 resolution dealt, in the main, with smaller populations, and with less and less dispute that independence was to be the goal, though it was not until 1974 that a domestic revolution in Portugal set her extensive empire on the road to it. By the time the sea-bed question was brought before it, the attitude that the General Assembly would take on any question which was seen as a dispute between European colonial powers and their subjects was a foregone conclusion. Equally predictable, if it came to a vote, would be the General Assembly's stance on an even broader domain of controversy, that of economic relations between the world's rich and the world's poor. The United Nations Conference on Trade and Development (UNCTAD) had, in 1964, confirmed the emergence of an alliance of countries whose common interest, as they perceived it, lay in the fact that they were not developed, the Group of 77. The resolutions adopted at the session illustrate the platform which an overwhelming majority of the members of the UN would support. In 1967, then, the General Assembly stood high in the opinions of developing countries as an organisation that could be made to sympathise with, and formally endorse, their aspirations; for many Western developed states, and especially those which had been colonial powers, experience had taught them to see it as a dangerous meddler, to be kept at arm's length; but others, including the USA, retained a vestigial, if eroded, confidence in it as a potential Cold War ally. The socialist states, on the other hand, had had bitter experience at its hand, which pointed almost unequivocally, and coincidentally with their Marxist ideology, at distrust of any international organisations in which socialist states were not in a majority; and consequent unwillingness to see them, and in particular the United Nations General Assembly, given any independent powers.

The Soviet Union thus preferred to deal with other major powers directly rather than through a United Nations forum, or at least to have approached such a forum only after reaching some agreement with these other parties directly concerned. In fact, before the Pardo initiative, the USA and the USSR, at the latter's initiative, had been preparing to convene a law of the sea conference on the territorial sea, fishing limits, and passage through straits. (6)

Those states, then, that might expect, in the not too distant future, to be able to take part, either by themselves or through their companies, in the mining of the sea-bed, had no reason, in 1967, to view the General Assembly as sympathetic to their interests. The poor "South" could already overwhelm them numerically. The steady influx of new members after that date, almost all from the Third World, accentuated this situation. As a result, the mining states increasingly looked for modes of negotiation which, if they did not rule out such majorities altogether, at least set them at a comfortable distance.

From 1967 to 1973 the General Assembly considered the item with the help of a body created especially for it, the Sea-Bed Committee (here sometimes abbreviated to SBC) (7). The Committee, originally consisting of 35 states, was increased to 42, and reconstituted, in 1968, as a standing body, with a promise that one-third of each regional group would be rotated, in membership, every two years. In fact, this did not happen. In 1970, the Committee was simply enlarged by the addition of a further 44 states, (8) and in the next year, 1971, following the replacement of the Kuomintang by the People's Republic of China in the United Nations, the latter, with four other states, was added to make a total membership of 91. The Committee met three times each in 1968 and 1969 (with additional meetings during the 1969 General Assembly), and twice in each of 1970, 1971, 1972 and 1973. With one exception (the third session of the AHSBC in 1968, which was held at Rio), all these sessions were held at New York or Geneva (as were all sessions of the Conference, except the Caracas session of 1974).

Not only did the Sea-bed Committee's composition change between 1967 and 1973, its mandate and structure changed also. The mandate given to the Ad Hoc Sea-bed Committee in 1967 emphasised "study" and suggested open-ended consideration of the peaceful uses of the sea-bed "beyond the limits of national jurisdiction"; the only request for advice it contained was vague and tentative in the extreme. It asked that the Committee's study should include

"an indication regarding practical means to promote international cooperation in the exploration, conservation and use of the sea-bed and the ocean floor, and the subsoil thereof, as contemplated in the title of this item, and of their resources, having regard to the views expressed and the suggestions put forward by Member States..."(9)

By 1968, the Sea-bed Committee was being requested to study "the elaboration of the legal principles and norms which would promote international cooperation in the exploration and use of the sea-bed and the ocean floor" (etc.), and "the ways and means of promoting the exploration and use of the resources of this area ... bearing in mind the fact (sic) that such exploitation should benefit mankind as a whole"

59

and to make recommendations to the General Assembly on these and other matters. (10)

In the next year, the General Assembly, after commending the SBC's work so far, called upon it to submit a draft declaration of principles in 1970. At the same time, it also set in motion the process whereby a whole range of other law of the sea issues might come under scrutiny. Malta had asked for a conference to be called specifically on the limits and regime of the sea-bed. The General Assembly, instead, decided (11) to ask the Secretary-General to canvass the opinions of member states on the holding of a general conference.

It was in 1970 that the committee's mandate underwent its biggest change. The General Assembly, on the same day that it adopted the Declaration of Principles, also decided to convene, by 1973, a general conference of the kind it had canvassed the previous year. The issue raised by Pardo, now defined as

"the establishment of an equitable international regime — including an international machinery — for the area and the resources of the sea-bed and the ocean floor, and the subsoil thereof, beyond the limits of national jurisdiction"

and "a precise definition of the area"

was indeed to be one of the conference's main tasks. But it was also charged with

"a broad range of related issues including those concerning the regimes of the high seas, the continental shelf, the territorial sea including the question of its breadth and the question of international straits and contiguous zones, fishing and conservation of the living resources of the high seas (including the question of the preferential rights of coastal States), the preservation of the marine environment (including inter alia, the prevention of pollution) and scientific research". (12) The Sea-bed Committee was asked to prepare draft articles for the former and "a comprehensive list" of the latter, as well as draft articles on them. (13) In other words, it was to become the preparatory committee for the conference as a whole.

This change of mandate naturally affected the SBC's internal structure. Until then, like the AHSBC that had preceded it, it had had two subsidiary bodies, the "Legal" and the "Economic and Technical" Sub-Committees (entitled "Working Groups" in the AHSBC), with the addition of a selective and somewhat controversial "Informal Working Group" in 1969. (14) Now that the General Assembly had decided to hold a comprehensive conference, the Sea-bed Committee, at its next meeting on March 1st 1971, ran into difficulty as to how it should proceed. It took the chairman until March 12 to prepare the ground for an acceptable solution, whereby three entirely new sub-committees were created, one (Sub-Committee I) charged with what had originally been the SBC's sole concern, the elaboration of a regime and machinery for the international area of the sea-bed; one (Sub-Committee III) concerned with the preservation of the marine environment (including the prevention of pollution) and scientific research, and one (Sub-Committee II) for all other issues, including that of defining the limits of coastal state jurisdiction over the sea-bed and thus the boundary of the international area. These three subcommittees, with this allocation of tasks, continued until 1973 when they became the three main committees of the conference.

For these six years, from 1967 to 1973, the pattern of decision-making alternated regularly between the Sea-bed Committee itself, which proceeded only by consensus, and the First Committee, and Plenary, of the General Assembly, in which, of course, majorities could outvote minorities. In fact, even in the General Assembly, the numerical majority, with one conspicuous exception, used its power sparingly. Of the eighteen resolutions passed between 1967 and 1973 on the issue, only five were opposed, and of those four dealt with procedure, (i.e., the studies the Secretary-General was to be asked to make, the kind of conference that should be explored, and the membership of the committee itself). The exception was 2574D of 15 December 1969, the 'Moratorium' Resolution, where, by little more than the minimum two-thirds majority required (62 to 28 with 28 abstentions), and against the opposition of all the most powerful states of the world who were then in the UN, the Assembly declared that "pending the establishment of the ... international regime",

"States and persons, physical or juridical, are bound to refrain from all activities of exploration" and that

"No claim to any part of that area and it resources shall be recognised".

The resolution polarised the United Nations at the time, and its validity has never been accepted by the outvoted minority. Yet the gist of it was very largely embodied in the unopposed Declaration of Principles of the following year, one of which was that "All activities regarding the exploration of the resources of the area and other related activities shall be governed by the international regime to be established" (para. 4). Neither resolution defined the limits of the area, or attempted to freeze existing limits to coastal state jurisdiction.

Some of the most important decisions affecting the form of the negotiations were taken at the 1973 General Assembly, after the Sea-bed Committee had met, as such, for the last time. First, it was decided who should be invited to the conference. The aim was generally agreed to be universality, and China, the socialist states of Eastern Europe, and some others, strongly supported a proposal that "all States" be invited. In opposition to this, the West favoured the "Vienna formula" which would have limited the participants to "members of the United Nations, or of its specialised agencies, or of the IAEA, or States Parties to the ICJ Statute" (Oda 1977, p.318), since the specialised agencies included in their membership Western orientated non-members of the UN, like Switzerland, the Republic of Korea and the then Republic of Vietnam, but not the Democratic Republic of Vietnam" (i.e. North Vietnam). The idea of inviting "all states", though attractive in principle, would have put on the Secretariat the impossible burden of deciding which entities, other than the UN's own members, were states. (Was Ian Smith's Rhodesia, for example?). The General Assembly resolved this by accepting the Vienna formula with additions: North Vietnam, Guinea-Bissau, and the UN Council on Namibia. A plea that South Vietnam (i.e. the Provisional Revolutionary Government of South Vietnam) should also be invited, was dropped, and as a result, in protest, North Vietnam refused to come to Caracas.

More importantly, the 1973 General Assembly helped to shape the negotiating process at the conference in two important ways. First it went a long way towards establishing consensus as its basic modus operandi. A "gentlemen's agreement", read out by the chairman of the Sea-bed Committee on 26 October, made plain the intention to devise procedures to ensure that no votes were taken at the conference until all efforts at consensus had been exhausted. Secondly, it committed the conference, for the first time, to reach a single comprehensive convention, rather than a series of separate conventions on different issues.

Finally this same General Assembly, in the wake of Pinochet's coup, and the overthrow of Allende, in Chile, decided to transfer the Second Session - the first to deal with substance - from Santiago to Caracas, and to hold the first session, limited to procedural matters and to not more than two weeks, at New York, when the Assembly had concluded its own business. This meant that the target the General Assembly had set itself in 1970, of holding UNCLOS III in 1973, could be notionally fulfilled; but the real beginning of UNCLOS III was not until Caracas. In 1973 it was foreseen that one session, even of ten weeks, as it had now become, might not be enough to conclude the conference's business, and provision was made for a further session in Vienna in 1975 (subsequently, of course, transferred to Geneva). Such was the hopeful spirit in which the General

Assembly took its final decisions in 1973, prior to the opening of the conference itself.

The chief result of this six-year interval between Pardo's original proposal and the convening of UNCLOS III was probably the growth of the idea of a comprehensive generally-accepted convention as the goal of the conference. This was remarkably ambitious. It certainly made sense to seek a generally-accepted text, rather than one in which a perhaps powerful minority had been outvoted; that made it more likely that the outcome would actually provide a framework for regulating marine activities. Nevertheless, the task was formidable, even if it had dealt only with the common heritage of mankind. To make it also comprehensive, covering in effect all aspects of the law of the sea, was to enlarge that task to mountainous proportions. Yet in embarking on this monumental commitment the widely expressed thinking at the time was that not only would it preclude the chaotic legal position that had followed the adoption of four separate conventions by UNCLOS I, allowing states to ratify the conventions they benefited from and simultaneously reject norms embodied in the others; it would actually make agreement easier, by permitting a state unhappy with some aspect of the convention to be compensated, within the overall "package deal", by concessions on other points.

The stakes of UNCLOS III were already high, but these two tendencies substantially increased them. The difference between success or failure now became, not merely the difference between exploitation of the deep sea-bed (at any rate) being regulated by an international body, entrusted with ensuring that mankind as a whole would benefit, and exploitation under competing national auspices; it could also mean, in large part, the difference between order and chaos in the law of the sea as a whole. Not total order, and total chaos: that would be an over-simplification. But the very fact of having a general conference in this field reinforced the state of flux into which the law, at that time, had fallen, and thus added to the obstacles which had to be surmounted, if not by substantive agreement, at least by some form of words, before the sea-bed itself could be satisfactorily disposed of. With hindsight, the extremely slow pace of negotiations, though not generally anticipated at the time, is easy to understand.

The delays imposed by the way the question was handled from 1967 to 1973, and the fact that the regime for the international area could be the subject of such protracted discussion, without regard to its boundaries, probably helped, as we shall see, the process whereby the boundary between "national" and "international" sea-bed was virtually settled by the consolidation of irresistible coalitions elsewhere. (15) At the least, it allowed states, without embarrassment, to proclaim their adherence to the "common heritage" principle, while advocating extensions of coastal state jurisdiction at its expense, or refusing to discuss the latter question. This schizophrenia was encouraged by the division of functions among the sub-committees of the SBC and the main committees of the conference itself, where, as we have seen, consideration of the _regime_ and _machinery_ for the international area fell to Sub-Committee I, and consideration of its _extent_ to Sub-Committee II. This division was perpetuated into the main committees of the conference itself.

A further effect of the practices of the SBC lay in the increasing formalisation of its work, which made the allocation of chairmanships and other offices into politically delicate tasks, constrained by the need to balance the gains of the five main regions, (Africa, Asia, Latin America, Eastern Europe and "Western Europe and others"). The appointment of Paul Engo of Cameroun to the chairmanship of Subcommittee I in 1972 conformed to this rule. Much of the work of the committee, from that time until 1975, was done by a working group under the chairmanship of Christopher Pinto, of Sri Lanka. The functions, and size, of this working group steadily increased, until it became, in effect, the committee itself meeting "informally" under a different name; but there could be no question of Engo being replaced by Pinto, since the Sea-bed Committee's chairman, H. Shirley Amerasinghe, who went on to become conference president, was also from Sri Lanka. No one continent, let alone one state, could be permitted two such important chairmanships. As a result, whatever the working group achieved had to be seen as tentative, unofficial, and subordinate to the sub-committee and its official chairman.

One final way in which the Sea-bed Committee's deliberations may well have influenced the process of subsequent negotiations was through its membership; which, as we have seen in the last chapter, divided the participants in UNCLOS III into three groups of states, roughly equal in numbers: those appointed to the Sea-bed Committee in 1968 (and, for the most part, also to the AHSBC); those who joined this core of states when the SBC was enlarged in 1970 and 1971; and those who had had little opportunity to familiarise themselves with the subject before UNCLOS III, including some non-members of the UN, like Switzerland, and very new members, like East and West Germany (admitted 1973). The fact that the veterans included proportionately more developed states and fewer land-locked than were eventually to participate in UNCLOS III, favouring those with the capacity to act in ocean politics, may have helped to reduce the divergence between expressed aspirations, and the "realities" of power politics.

This process of twice, (16) in effect, starting again with a large influx of newcomers, though it condemned the proceedings to tedious doses of repetition, as each new influx went on to raise already familiar issues, may also have made for continuity and a sense of direction. Slow though it was, there was a consecutive line of progress, on the question of the regime and machinery, from the introduction of Pardo's item in 1967, through the Declaration of Principles of 1970, to the First Committee's deliberations in UNCLOS III. More and more of the framework of a regime was agreed. It is possible that the General Assembly's decision to entrust the question to a committee of the size it did in 1967 maximised the chances of producing a coherent, internationally-authorised, system of exploitation. Had it immediately called a conference of all members (and the outsiders) it might have produced no agreement at all, with perhaps a system adopted by majority vote; had it entrusted the question to a smaller committee, that committee might have found it easier to agree, but might then have been overruled by a more numerous body. As it happened, the influx of new members was never overwhelming, although the membership was slightly more than doubled in 1979; and though the process of successive enlargement always had some revolutionary potential, there was at no point a drastic reversal of preceding trends.

The six years of Sea-bed Committee meetings, punctuated by annual debates in the General Assembly and the First Committee, did not do enough to enable UNCLOS III to get down at once to the substance of its business. Its first, short, session, in December 1973, was entirely given over to procedure, and did not complete its task. The drafting of the conference's Rules of Procedure took a further week, the first of Caracas's ten-week session. When, at the end of it, virtually universal agreement was reached (the only discordant notes being mildly expressed objections, from diametrically-opposed standpoints, by Peru and the Soviet Union), there was much euphoria, and President Amerasinghe won great acclaim for his part in bringing it about.(17)

What New York had not settled, and Caracas did, was, at bottom, the question of what would ultimately count as a conference decision. Put bluntly, if it came to a showdown, what kinds of majorities, in what circumstances, could override what kinds of minorities, and how? The answer given by these rules was that, before any vote was taken in plenary session on a matter of substance, any fifteen delegates, or the president at his discretion, could invoke a delay of not more than ten calendar days, and that, when this had expired, or if it had not been invoked, the Conference could proceed to a vote only if it had previously determined, by a two-thirds majority, including a majority of the states participating in that session, that all efforts at reaching general agreement had been exhausted. A similar majority would be required for the original proposal then to be adopted. These rules were also to apply to the eventual decision to adopt a convention, except that the "cooling-off" provisions were to be replaced by a delay of four days between the last vote on any article of the text and the putting to the vote of the convention as a whole. In the three main committees, analogous but less formidable obstacles were put in the way of voting. Any fifteen members could impose a maximum delay of five days, but the committee would, in the end, vote by a simple majority. Simple majorities were also to apply to procedural decisions in plenary, including challenges to a presidential ruling that a question was procedural. (18)

Ed Miles, as we have seen, saw these rules as contributing to the "extremely slow pace of the conference", (and he was referring only to the Caracas session!). In his view, "the fight over the draft rules was extremely important because it imposed on the conference a most unwieldy method of making decisions ... the rules adopted contributed to delays by enshrining the consensus procedure at the heart of the conference and by making voting so difficult and potentially time-consuming that even without contention it would have been difficult to complete the agenda in ten weeks (Christy, Clingan, Gamble, Knight and Miles 1975, pp.46-47). Moreover, the fact of their adoption was "convention-breaking" in that they "constituted a repudiation of the normal General Assembly Rules of Procedure".

This was certainly an innovation, with no exact precedent in UN conferences, and in sharp contrast to UNCLOS I and II, in 1958 and 1960; but it can hardly be argued that it has, at any time, prevented UNCLOS III from reaching a workable outcome. The only outcome it may have prevented is one in which a minority would have been outvoted by a majority in no position to implement its decisions. Even in the hypothetical event of there being "no contention" - and this is totally unreal, since there was scarcely any point of the conference's agenda that was contention-free - the procedures would not then have been invoked. They have never been used to delay the conference when there has been general agreement on how to proceed.

The fact that voting was ruled out as a routine method of decision-making reinforced the necessity of constructing a "package deal", a convention that, taken as a whole, left no significant state, or group of states, so seriously dissatisfied as to prefer to stay outside it, and disregard its norms. The commitment to establishing a sea-bed authority promised to be an important part of this overall package. To those states that might otherwise have had no role in ocean mining it offered a share in regulating exploitation, a share, likely to be relatively greater for developing countries, in the revenues therefrom, and possibly (and this perhaps was for developing countries the biggest attraction) participation in the process of exploitation itself. There would need, however, to be a package within the sea-bed authority negotiations themselves. Prospective exploiters, whether companies or states, were being asked to accept restrictions on what they could do and how they could do it, and to pay what, in effect, were taxes to the Authority. They would insist, in return, on some assurance of being granted an internationally-validated title at a not intolerable cost. Thus even within the realm of First Committee matters, taken in isolation, there were constraints which narrowed the range of potentially acceptable outcomes. There were also constraints on what might emerge from the other committees. In particular, extensions of coastal state jurisdiction had somehow to be reconciled with the concerns of maritime and naval powers, particularly on the question of straits, and the frustrations of the land-locked, and of their geograph-ically-disadvantaged allies, had to be sufficiently assuaged to persuade them not to vote against the convention en masse. To some extent, disappointments in one sphere could be set against gains in another, particularly, for non-ocean mining states, gains from the establishment of a sea-bed authority; but many states had their core interests, in some sphere or another, which would rule out, for them, acceptance of any convention which jeopardised such interests, whatever other attractions it might hold.

TECHNIQUES OF MULTILATERAL DIPLOMACY

Public Debate

One of the most important functions of any major multilateral conference is to allow participants the opportunity of publicly expressing their views on the issues before it. The historic character of UNCLOS III reinforced the inclination, to which all states priding themselves on an at least notional independence were prone, to enunciate and justify their position.

The Caracas session spent two and a half weeks, in plenary, hearing general statements from 115 states, (19) eight international organisations and one non-governmental organisation. Those of the states were often delivered by political figures, such as the relevant ministers, attending for brief visits, rather than the professional diplomats who in practice normally led national delegations. In addition, and beginning before addresses to the plenary had been concluded, each committee offered every member the opportunity to speak generally on the issues before it. Thus the First Committee heard sixty-five states speak to it in the first five days of its own general debate.

These were not, of course, debates in the sense of an intellectually coherent exchange of argument and counter-argument. It was rare, particularly in the plenary, for anyone to refer to a previous speaker. There were even inconsistencies between what was said on behalf of a delegation in plenary and what was said on its behalf by a different delegate in a committee a few days later. Nevertheless, such debates, and the kind of reception each statement received, were instructive, certainly for observers, and, not infrequently, for the delegates themselves. The 115 speeches in plenary at Caracas demonstrated the range of interests of states and gave some idea of what they were prepared to tolerate. It was clear thereafter, as it had not been before, that some sort of a 200-mile economic zone would be part of any convention to emerge: it was also clear that some embodiment of the idea of the sea-bed as the common heritage of mankind, already vindicated by the 1970 Declaration of Principles, would also be an indispensable part of such a convention. More disputably, the airing of the views of the land-locked and geographically-disadvantaged states may have established them as a category which could not be entirely ignored, in spite (or because) of their acceptance in principle of the 200 miles economic zone.

After this initial burst of statements public discussions became much less common. Thereafter, public debates tended to be confined either to end-of-session reviews, or to comments on proposals that had been officially put before the conference. For instance, in the New York part of the ninth session in 1980, four plenary meetings on two consecutive days (April 2nd and 3rd) were devoted to responses to a proposed second revision of the Informal Consolidated Negotiating Text. (20) Open meetings were also used by chairmen of committees or groups to report progress or lack of it; and where procedural changes were launched, to give delegates a chance, usually taken only by the representatives of regional and other groups, to proclaim their attitude towards them. Some of these were quite brief and uncontroversial.

Alternatives to Public Debate

It was not surprising that public debate should become the exception rather than the rule. Conferences subsist on a variety of diets, and it is the mixture, rather than any one staple food, that nourishes or debilitates them. In the case of UNCLOS III, the diet had three main other ingredients: according the initiative to the president, or to the chairman of one of its committees; working through more or less select groups of states in which each major region or relevant category is deemed to be represented by two or three of its number; and informal meetings of the whole.

This last procedure has been a favourite one, particularly in the First Committee. Though open to all states, such meetings have been closed, not only to the press, but also to observers from non-governmental organisations. No official records of them are published, though tapes are apparently kept by the secretariat, which permits each delegate to check only what he himself has said. The gist of their proceedings is reported by their chairman to public meetings of the same bodies. At Caracas, for instance, the First Committee, after five days of public debate, went into "informal meetings of the whole" under the chairmanship of Christopher Pinto of Sri Lanka, for most of the remainder of the session, the culmination of the trend referred to earlier. Its mandate had originally been confined to the "international regime", that is the set of legal principles that were to govern exploitation of the international area; but, by Caracas, it had come to include in its purview the structure of the authority, and the question of who should exploit the area, and to delegate other, less controversial, items to a smaller group within it (Christy, Clingan, Gamble, Knight and Miles 1975, p.58).

This "committee of the whole" continued, under Pinto's chairmanship, up to the end of the 1975 Geneva session. An attempt towards the close of the Caracas session to set up a new selective, but "open-ended", working group of fifty, again under Pinto, to operate between sessions, was first unanimously accepted and then, at a later meeting, so vehemently criticised that it was dropped. The Geneva session ended with some tension between Pinto and Engo, and Pinto's group as such was disbanded. This did not prevent the First Committee from meeting informally, as a whole, in the fourth session in the Spring of 1976; such meetings were frequent, but now Engo chaired them, as well as the official ones. The substantively barren fifth session, also in 1976, tried out several procedural innovations; but the next, in 1977, reverted to private meetings of the whole, mostly in the guise of a "chairman's working group" headed by Jens Evensen, of Norway. From 1978 on, the emphasis was shifted from the main committees to a set of seven "negotiating groups", each assigned one of the "core issues", as yet unresolved, from the First and Second Committees. The chairman of the three from the First Committee were Frank Njenga of Kenya, Tommy Koh of Singapore, and Engo himself. Negotiating Group 6, chaired by the Second Committee's own chairman, Andres Aguilar, was also relevant to the "common heritage" since it dealt with the limits to the shelf, and the obligations of coastal states to share revenues with the authority. All these negotiating groups, like the committees from which they derived, were, again, in effect, informal meetings of the whole.

The prevalence of such private meetings of the whole suggests, surprisingly, that states attending the conference (in effect, every state in the world), in spite of their apparently fundamental cleavages, have something in common with each other which they do not share with the general public or with those non-governmental organisations to whom they concede the privilege of observer status. The explanation may lie in the diplomatic professionalism of delegates. Diplomats seem readier to admit, before an audience confined to their peers, that they do not understand some technical aspect of the debate. They trust each other not to exploit their personal limitations. There is indeed, it seems, an "esprit de corps diplomatique". Further, in private meetings, diplomatic courtesies can be attenuated. A brusque rejection of a proposal is less like to precipitate a showdown, or preclude a subsequent volte-face. In spite of such considerations, the fact that UNCLOS III, with its extreme diversity of membership, has so often had recourse to private meetings of the whole, remains a telling one.

By contrast, attempts to delegate the business of Committee I to more select groups have generally come in for some sharp criticism from the unselected. As we have seen, the idea of continuing negotiations, after Caracas, and earlier, through Pinto's "open-ended" working group of fifty, ran into strong opposition. Paul Engo's attempts at the fourth session to work through a number of small, though again supposedly "open-ended", groups on specific topics, (as well as through a committee of the whole) led to charges that he was running a "First Committee mafia"; and certainly produced no firm advances. The fifth session experimented with, first, an informal "workshop" with two co-chairmen (from India and the Netherlands), which was in effect a committee of the whole, and then an "Ad Hoc Negotiating Group" of twenty-six, which all other delegations could attend but not address. These, again, had no immediate yield: but the principle incorporated in the latter device of an "arena" within which a few delegations would do the talking and the rest could see what was going on, was seen by some as the key to success, the only manageable and effective compromise between 'oligarchy' and 'democracy'. For the next two sessions, the official business of the First Committee was left to committees of the whole, but at the Geneva part of the eighth session, in 1979, a resolute attempt was made to channel it through a new body, apparently more select than any of its predecessors, the Group of 21, which was to consist of ten developing countries, eight Western developed states, two Eastern European socialist states and China. Apart from China, each was supposed to represent its group or regions, and delegations from other countries were not even to be permitted to observe. The choice of representatives created acute tensions within several of the groups, which was only partially relieved by the designation of "alternates" for each representative, thus increasing the effective

70

membership well above the original twenty-one. Resentment still ran high among those excluded, while _within_ the Group of 21, the early meetings were leaden and totally discouraging. Within days, the rules were relaxed to permit all delegations to attend even though only twenty-one, at any given time, could actively participate. This return to the "arena" principle of the fifth session was vindicated by the quite remarkable progress that characterised the closing stages of that session.

The major procedural innovation of UNCLOS III, however, was the idea of asking committee chairmen to product "negotiating texts", that is, to articulate proposals, on which subsequent negotiations could be based, in the form of published, and more or less comprehensive, drafts of the articles with which their committees were respectively concerned. It was at the third session, at Geneva in 1975, that this device was first adopted, at the suggestion of President Amerasinghe. Hitherto the conference, and its three committees, had before them only proposals emanating from states, or groups of states. The nearest thing to a precedent, in its official proceedings, was to be found in the reviews which Andres Aguilar, the Venezuelan chairman of the Second Committee, had produced at Caracas of what he saw as the "main trends" in negotiations on each item. But Aguilar's "main trends", though not uncontroversial, listed on each contested issue several alternative positions. The idea of the single negotiating texts was that each should offer, on the matters with which it dealt, only a single formulation. In this respect, the obvious forerunner was the modus operandi of the fairly widely representative informal group which had met at Caracas, and earlier, under the chairmanship of Jens Evensen. That too, had worked though sequences of chairman's proposals; but these, of course, had no official status, were not published as such, and became important only when they elicited widespread agreement; whereas the single negotiating texts (SNTs), although they did not bind anyone, and their allegedly "informal" character was emphasised, _were_ published and thus became a focus of attention, going well beyond the confines of the conference itself. Those who, on a given point, had the SNTs on their side had not only a debating advantage, in that they could lay claim to "reasonableness" in accepting the officially suggested "compromise"; but also inertia, in that time might permit the effective challenging only of those parts of the SNT with which one or other participant glaringly disagreed. The status of the SNTs was demonstrated in 1976, when the fourth session rejected, in effect, what President Amerasinghe had published as Part IV of the SNT, on the settlement of disputes, on the grounds that, though based to a considerable extent on the work of an informal group on that subject, it had not, unlike the other three Parts, been preceded by a full debate in which all delegations had had the chance to speak, and so did not carry the same weight. Thus it was only after the setting up of an informal plenary, also known as the 'fourth committee', at the same session, that the president, as its chairman, was deemed entitled to issue what could properly be called an SNT Part IV.

As a device the SNTs seemed both apposite and fruitful. Revised single negotiating texts (RSNTs) followed, in the case of Parts I, II and III, at the end of the fourth session and for Part IV, at the end of the fifth. At the sixth session, in 1977, (or rather, shortly after it finished), the first attempt was made to produce a model of the promised comprehensive convention to which the conference was pledged, in the form of the Informal Composite Negotiating Text (ICNT). The fact that, in each case, the text was issued at the very end of, or after, the session, prevented, and was intended to prevent, immediate public objection being taken at the session itself.

These texts, inevitably, put great weight on each committee chairman, individually, for matters that concerned his committee. Each was wholly responsible for his "part" of the SNT and RSNT, and even with the ICNT, the "collegiate" principle, by which it was drafted, meant that each had the last word on anything in it that fell solely within his committee's mandate. Thus, as the texts progressed, the voice of the chairman became notionally the voice of the conference as a whole. In Committees II and III this seems to have been wholly conducive to agreement; but in Committee I it was not. Each of Paul Engo's Texts — the SNT, the RSNT, and his part of the ICNT - was repudiated by one or other side of the great divide in that committee, and in each case severe criticism was levelled not just at the substance of the text, but as the process by which it had been reached.

In two of these cases – the SNT and the ICNT – the charge against Engo was that his text did not reflect the trend of negotiations, which had been largely conducted, in each case, by someone else. In 1975 the someone·else was Pinto, who, as we have seen, had played a very prominent part in first subcommittee, and subsequently First Committee, discussions ever since 1972. After intensive consultations among all the major interest groups, Pinto, as chairman of the working group of the First Committee, produced a full draft of a single negotiating text, which he eventually submitted, as yet unpublished, to Engo. (21) Engo, though following Pinto's text on most points, introduced a considerable number of amendments, which persistently tended to favour the position of the Group of 77, much to the indignation of the industrialised countries, whose delegates had felt that, in their discussions with Pinto, their view had at least been listened to and understood. In 1977 this pattern was repeated, perhaps even more momentously. In the previous February, more than eighty heads of delegation had attended inter-sessional talks in Geneva, at the initiative of the Norwegian Ambassador, Jens Evensen, and elements of a possible compromise had been sketched. There was therefore wide support for, and high hope of, the "chairman's working group" (22) under Evensen, established in the sixth session, as a mode of conducting First Committee business. This allowed Evensen to elaborate these sketches, and put them through a series of revisions in the light of comments from one or other quarter, reaching a point at which he believed that, "by a miracle", he had achieved consensus on many (though not all) of the First Committee's concerns. But, again, the "informal" proposals were not allowed to go into the ICNT as they stood. Engo, claiming that they did not reflect consensus, made more than a dozen significant alterations in incorporating them.

In the case of the RSNT, in 1976, in contrast to the SNT and ICNT, it was the Group of 77 that objected to Engo's text. The charge here was not that the RSNT did not flow from the negotiations at that session, but that those negotiations had been confined to too narrow a circle of states. This circle had certainly included developing states, among them Peru, which held the chairmanship of the Group of 77 in the First Committee; but, even with Engo's help they failed to carry their rank and file.

The starting point of the negotiations had been the strong dissatisfaction of the United States, in common with other Western powers, with Part I of the SNT. In December 1975 the United States had privately submitted to Engo a bulky volume of proposed amendments to that text, which would have left scarcely an article of the original unscathed. Intensive inter-sessional negotiations, and some striking and covert diplomacy between the United States and the major land-based producers among the developing countries, had prepared the ground for revising the SNT at the session itself. Much of the session was taken up, in the First Committee, with business-like and harmonious discussion of short papers, each dealing with one or two articles, or paragraphs of the 'basic conditions of exploitation', produced by Engo, and thus called PBEs. These would have revised the text in direction of, but (in most cases) falling some considerable distance short of, the amendments the US had proposed. Where alternative formulations emerged from the floor, the delegates concerned were asked to constitute informal negotiating groups and reach a compromise through private meetings. The process was not confined to the points of major controversy but extended also to less controversial features of the text. At the time, the whole process seemed constructive and promising - so much so that the United States, whose Republican Administration was widely (and correctly) expected to lose the Presidential elections that year, pressed for, and persuaded the conference to accept, a further session in 1976 to complete the job. Engo's RSNT, which tended, on the points discussed, to follow his PBEs with such amendments as had seemed to find general favour, was expected to be the basis for subsequent negotiations.

It therefore came as something of a surprise, at least to the general public, that, at the beginning of the fifth session, a succession of developing country speakers should emphatically declare that no part of the RSNT Part I could be regarded as agreed, and should explicitly denounce the processes of small meetings and informal diplomacy that had brought it about. To the delegations themselves, it may have been less surprising. It was subsequently revealed (23) that, even at the fourth session, the Group of 77, having been given some indication of what the text might contain, had written to Engo reserving its position, asking for its own views to be taken into consideration, and expressing the feeling that, pending study of the revision, the SNT was an adequate basis for discussions; and there were even rumours that the US delegation, which pressed so strongly for a second session in 1976, were not, by the time that it opened, expecting to succeed. (24)

The crisis of 1976, unlike those of 1975 and 1977, was a fundamental one, and can hardly be blamed either on the theory or the practice of chairmen's texts as negotiating devices. The innovations to which it gave rise have already been looked at. In the short-run they failed. Delegates comforted themselves with the reflection that the persisting deadlock represented a necessary stage in the negotiating process, confronting them for the first time with the stark possibility that the conference would fail, because of the adherence of the developed states, and the resistance of the developing, to the principle of a dual system. The details of that debate will be recounted in Chapter 7. In terms of the flow of negotiations, the view that the conference had to go through the barren intermission of the fifth session, though expressed only in retrospect, may well be right. There is however, another interpretation. The impending American Presidential election and expected defeat of the Republicans may have led the Group of 77 to hope that they could win more out of Jimmy Carter and (as it turned out) Elliot Richardson, than they had from Kissinger and Leigh Ratiner (though in fact the latter pair had made some substantial concessions to developing countries' concerns). On this view, the conference might have saved the expense of the fifth session without loss of momentum.

Whatever the truth of that, the fifth session ended in a sober mood, and, in its arrangements for its successor, made one change of some importance. Recognising that the main obstacle to a treaty now lay in the First Committee, it decided that the first two weeks of the sixth session should be primarily focused on that committee's work and that participants should be represented in it by their heads of delegations. This sense of urgency was reinforced by Norway's proposal in the General Assembly, and the emergence of Evensen as the "honest broker", to which reference had already been made.

The high expectations and frustrating outcome of Committee
I negotiations in the sixth session led the new
Administration to order a review of the whole question of
United States' participation in UNCLOS III, and to change
from opposing, or at least delaying, unilateral legislation
to actively supporting it. The fact that Engo had produced
three texts, none of which was accepted "as a basis for
negotiation" by both the United States and the Group of 77,
suggested that, if the problem was soluble at all (as it had
seemed to be at the sixth, where the Group of 77 had at one
stage given qualified support to the Evensen proposals), it
required radically different techniques to solve it.

Mention has already been made of the remedy which the
conference now devised, the identification of seven core
areas of disagreement in the domains of Committees I and II
and the establishment of a separate negotiating group for
each. It took a crowded time-table of inter-sessional
meetings, some selective, some open to all participating
states, to achieve this. Each chairman was to report to the
president, as well as to his committee chairman, thus
diluting, in some measure, the latter's stranglehold on any
amendment of the official texts, although each committee
chairman had the discretion to require that proposals
emanating from a constituent negotiating group were
discussed in his committee before being considered by
plenary. (25)

The committee chairmen's powers were further limited by
the simultaneous adoption, by the conference, of a new mode
of revising the ICNT, whereby

"Any modifications or revisions to be made in the
Informal Composite Negotiating Text should emerge
from the negotiations themselves and should not be
introduced on the initiative of a single person,
whether it be the President or a Chairman of a
Committee, unless presented to the Plenary and
found, from the widespread and substantial support
prevailing in Plenary, to offer a substantially
improved prospect of a consensus". (26)

As far as the negotiation of a sea-bed authority was
concerned, the effect of these changes was to limit Engo's
discretion, to permit the negotiating group chairmen to work
out compromises that could be officially considered as they
stood, and not only insofar as he chose to incorporate them,
and to enable these texts to be considered, as possible
revisions to the ICNT, in the sessions in which they were
produced. These changes were salutary, but they created at
least one further problem. They tended to split the
negotiations over the regime and machinery for the common
heritage, which was already separated from consideration of
its geographical limits, into three further self-contained
compartments, and made any overall bargaining which cut

across these compartments difficult. There was need, it seemed, for a body in which the USA, or the Group of 77, could indicate, however delicately, that they would accept x in Negotiating Group 1, if they could get y in Negotiating Group 2. The two new chairmen, Njenga and Koh, succeeded in narrowing the gaps between positions in 1978 and the early part of the Geneva portion of the 1979 session; but the problem remained, and it was this that led to the creation, as already described (27) of the Group of 21, with Engo once more at the centre of First Committee negotiations.

This time it seemed to work. A series of amendments to the plenary, by Engo, were accepted as offering a "substantially-improved prospect of a consensus". The revision consisted of two parts. The first arose directly out of the work of the Group of 21. (28) Curiously, Engo, in his report to the plenary, said of the report of this group that it "in no way attempts to relate to a revision of the ICNT". The other basis for the revision was Engo's own report. The decision to accept such proposals as improvements to the ICNT was taken by the conference officers, to whom the plenary, in its last meeting of the Geneva part of the session, formally entrusted it. Further revisions were made at the end of each part of the ninth session. (29) The first was produced by roughly similar means. The chairmen of the three negotiating groups originally charged with Committee I questions produced texts which they had successively refined in the light of comments; and the group of legal experts, under Harry Wuensche, of East Germany, and a further group on production limitation (originally part of Njenga's brief) under Satya Nandan of Fiji, did likewise. They were then debated in the First Committee and the plenary, the latter debate encompassing 92 speakers over two days. (30) The officers then issued a revision of the ICNT, incorporating only changes already proposed in the latest texts of the various group chairmen, and then, with one exception, only if the plenary had shown that they had general support. The exception was the case, referred to earlier, of Article 162, para 2(i).

It is clear that in arriving at this revision the First Committee had moved far from what Engo now called "the debauchery of publicity and records". It had equally moved far from informal meetings of the whole. Instead, in the Group of 21, "for the first time, it was possible to have a limited number of speakers all accredited representatives of definite interests... this, clearly was an advance in the right direction and away from the unruly system of so-called open-ended meetings."(31)

Engo may bear some responsibility for the fact that it took so long to arrive at a conference text which could be accepted by developed and developing states alike, as a basis for negotiations; but in these expressions of opinion, he seemed to have correctly identified a process which the conference needed to discover before it could hope to pass through the eye of the needle on which its camel's humps had baulked for so long.

The one major element in Part XI of the ICNT which had not been satisfactorily resolved by the time the second revision had been issued was one which fell within Engo's own Negotiating Group 3. It concerned the important question of voting in the Council. In the resumed ninth session, in Geneva in the summer of 1980, a compromise was worked out among a small group of delegates representing the major industrialised states the USSR, and the Group of 77, together with Engo himself and Tommy Koh, who had earned considerable credit by his successful completion of the work of Negotiating Group 2, and was soon to be elected president of the conference in succession to Amerasinghe. This compromise was strongly criticised by the smaller industrialised states. They had been by-passed in its negotiation, and had been somewhat discriminated against in the existing provisions for the Council's composition, which the new deal did not change. Their dissatisfaction did not, however, extend to opposing its inclusion in a further revision of the text, and the collegium (the president, and the chairmen of the three main committees, and of the Drafting Committee, gave this third revision the title of Draft Convention. At that point, the main unresolved issue in the convention itself did not concern the "common heritage of mankind". It was the vexed question of the basis for the delimitation of the marine zones of jurisdiction, particularly the economic zone and the continental shelf, as between "opposite and adjacent" states. There remained, however, certain loose ends to be tied up with respect to the "common heritage", particularly the establishment of the Preparatory Commission to enable the Sea-bed Authority to come into existence as soon as the necessary ratifications were forthcoming, and the conclusion of transitional arrangements specifying how investment in sea-bed mining, undertaken under national authorisation prior to the Sea-bed Authority's birth, was to be reconciled with the system of the convention.

The Reagan Administration, however, by its abrupt announcement of a thorough-going review of its policy, coupled with an equally peremptory change in the personnel of its delegation, on the eve of the tenth session in 1981, was to shatter hopes that the achievement of a generally-accepted convention would require only the tying up of loose ends. Not only did the US delegation thereby exclude itself from substantive negotiations at the New York part of this session; but by making it clear that its review would not be completed until the end of the year, it ensured that even when the session resumed in Geneva in the summer such negotiations could still not take place.

Though the announcement was naturally greeted with great disappointment, and some indignation, particularly on the side of the Group of 77, the conference, on the whole, swallowed this bitter pill with impressive restraint. It did not entirely accept the US request to postpone formalisation of the text until 1982; and the Group of 77 vented its displeasure by refusing to discuss PIP (Preparatory Investment Protection) until the USA had come to a position; but when in August 1981 the text was formalised, time - three weeks - was reserved at the following session for further negotiations, before the invocation of Article 33 which was to signal the stage at which the text would be changed only by formal amendments. Moreover, the session proper was to be preceded by meetings of the Drafting Committee, which could also provide an opportunity for preparing the ground for such compromises as might be needed.

The 1982 session was to prove an exciting, though ultimately a disappointing, one. Its tasks could be put in two categories. There was the unfinished business of the conference, that is, matter on which the Draft Convention was silent. Essentially these were the three Ps: The question of participation in the convention (of national liberation movements such as the PLO, and international organisations such as the European Community); that of the establishment of a Preparatory Commission (Prep Com) to prepare for the establishment of the Sea-bed Authority between the signature of the convention and its coming into force; and "Preliminary Investment Protection" (PIP), a complicated but vital issue dealing with the conditions on which existing investors (or those claiming to be such), known as "pioneers", could be assured of the right to explore an area before the Authority had been constituted with a view to exploiting it, or part of it, within the convention when it at last came into force.

Then, secondly, there were the issues within the drafts of 1980 - 1981 which the United States wished to reopen.

The second of these tasks proved much more difficult than the first. On the three Ps, the conference went to work in a straightforward, constructive way, resulting in the now familiar sequence of proposals made by the conference president, or committee chairman where appropriate, and revised in the light of comments of delegations in the "general debate" and of private consultations, and then attached to the text, though in this case apart from "participation", not as elements of the convention but as resolutions to be adopted by the conference in conjunction with it, since by their very nature they had to be implemented before the convention came into force.(32)

Consideration of the US amendments to the draft convention, however, which were supported by six other sea-bed mining states (Belgium, France, Germany, Italy, Japan and the UK), meant undoing the agreements so painfully hammered out over the preceding four years (and in some cases more). Mediation could not easily come from the president or the chairman of the First Committee, or any officer of the conference, since the texts were theirs, and, (to elaborate) had no doubt won the support of certain states, in 1980, and perhaps also that of the Group of 77 as a whole, precisely because these same officers had been able to assure them that these provisions would have general support. Officially, the 1980 text was generally approved only as "an improvement on" previous revisions of the ICNT, not necessarily as acceptable in itself. In practice, though, it had been regarded as if it had been, apart from the delimitation question, the embodiment of agreement.

Such then were the four main devices employed by the conference, in an attempt to achieve the seemingly impossible task of writing a single comprehensive convention on which all, or virtually all, states could agree: public debate, informal meetings of the whole, more selective groups, and texts emanating from the various chairmen, and in particular, the official negotiating texts (SNT, RSNT, ICNT, and its three revisions) produced by the chairmen of the main committees. Each had its uses; each also, at times, hampered rather than furthered progress. UNCLOS III cannot, in respect of its procedures, be said to have hit upon a formula for assuring success; but in spite of its vast size and duration, it had been for the most part, innovative and flexible, particularly in first fostering and then, for the First Committee at least, curbing, the committee chairmen's initiatives.

The Presidency

One factor in determining how the conference was to proceed, which has so far been mentioned only in passing, is the quality and personal style of its president; and the presidency of Hamilton Shirley Amerasinghe lacked neither incident nor influence. He was in at the start. The Ad Hoc Sea-Bed Committee chose him as chairman at its first meeting in 1968. He had remained chairman of the Sea-bed Committee throughout its existence and then became president of the conference. Mention has already been made of the credit given him for the agreement on rules of procedure at Caracas, but the prestige which flowed from the success was diminished somewhat by narrowness of the vote by which, later in the same session, a ruling of his, to permit a fairly wide-ranging proposal by nine coastal states to be presented in plenary rather than in the relevant committees, was upheld (50 to 38 with 39 abstentions).(33) His suggestion, at Geneva, in 1975, of asking the committee chairman to publish single negotiating texts, had rescued the conference from stagnation, and was generally acclaimed;

yet though the idea proved acceptable, his own first attempt at an SNT Part IV, on dispute settlement, was as we have seen, widely criticised as premature. In 1977, when some states wanted to give him the power to overrule committee chairmen in drawing up to ICNT, others, particularly the coastal states, who firmly supported the chairman of Committee II, demurred, so that whatever he might thought of them, he lacked the power to prevent the Engo revisions to Evensen's texts being incorporated into the ICNT.

In general, then, UNCLOS III, in its earlier years, can be said to have accorded Amerasinghe a good deal of respect, but not, by any means, uncritical adulation. Nor had it relied exclusively on him or the other conference officers for procedural innovation.

The practice of intersessional meetings, which proved of some importance in 1977 and 1978, owed little to him. It was Evensen who had invited heads of delegation to Geneva in February 1977, which led in the First Committee to the chairman's working group and the Evensen texts; and the more crucial and complicated pattern of intersessionals in 1977-8, between the sixth and seventh sessions, which persuaded the United States to stay in the conference after the publication of the "totally unacceptable" ICNT, also originated in a private meeting, called by Evensen, of a "broadly representative" group of 28 countries, although when the ground had been cleared, "full" intersessionals were then called by Amerasinghe.(34)

The severest test of Amerasinghe's standing came at Geneva in 1978, after the new Sri Lankan government dropped him from their delegation. The Latin American states, led by Venezuela, insisted on the principle that no non-delegate could preside over the conference. The African and Asian states were equally determined to keep him in the chair. In the end, it took more than two weeks of the seventh session to decide, by a majority vote, followed by a temporary walk-out by the aggrieved minority, that he should be retained as chairman. This was the most dramatic split in the Group of 77 throughout the conference, and one of the rare occasions in which the Latin American states, in an international gathering, found themselves in the minority. Amerasinghe maintained great dignity and courtesy throughout this challenge to his personal position, but his capacity for influencing events seemed to be fatally weakened.

As chairman of the conference in its plenary session, Amerasinghe was known, and admired, for his dry good humour and fast gavel.(35) In apparent contradiction to his professed determination to search for consensus, he would push decision through with startling rapidity. This led, on occasion, to bewilderment, but rarely to resentment. He was generally careful to ensure that, in any matter of importance, the agreement of the main parties had already been secured.

Amerasinghe died suddenly in December 1980. Because of the principle of the division of the spoils among regional groups, his successor had to come from Asia, and three delegates from the regional island-states (Satya Nandan of Fiji, Tommy Koh of Singapore, and Christopher Pinto of Sri Lanka) competed for the job. All three, and particular Tommy Koh and Christopher Pinto, had made notable contributions to the progress of the conference. Pinto's work as chairman of the working group of the First Committee - not his only office in UNCLOS III - has already been mentioned. Tommy Koh had shown his healing touch at least as early as 1977 when he had reminded the First Committee, in the face of polarisation between the Group of 77 and the West, of the extent of its common ground. When three "hard core" issues that fell within the First Committee's preview were each assigned to a negotiating group in 1978, he became chairman of Negotiating Group 2, which inter alia settled the question of the financial obligations of contractors so effectively that even the Reagan review did not re-open the issue. In the summer of 1980 he was one of a small group of delegates who worked out the then generally accepted compromise on voting in the Authority's Council. Given the harsh divergence between the Reagan Administration and the majority of the conference which coincided with his time in office, he did not have much chance of crowning this earlier work with the adoption of a convention by consensus. His unassuming informality and sustained openness to new ideas enhanced what chance there was. It is difficult to imagine that in such a predicament, any other president could have done better.

Groupings

The features of UNCLOS III, and of its predecessor, the Sea-bed Committee, so far examined have been wished upon them, either by themselves, or by their parent body, the General Assembly, in official decisions. But other categories of factors, both inside and outside the conference, helped to shape its progress. Had they been otherwise, these same interests, earlier described, might have blended, and contended, in different ways and with different results.

Chief among the factors internal to the conference were the groups in which the multitude of delegations more or less spontaneously associated.(36) The conference took note of the existence of many such groups, afforded them rooms to meet, and announced and displayed their times of meeting, even though whether anything emerged from a group in the form of statements or proposals was left entirely to that group. These groups were of two kinds: already existing groups, like the regional groupings and the Group of 77, and groups generated by the issues before the conference, like the Territorialist group and the group of "Land-Locked and Geographically-Disadvantaged States" (LLGDS).

Conference-generated groups, though they have no previous history, can still generate their own momentum. The longer the conference lasted, the more continuity they enjoyed, and it may well be that some of them will persist, to further the interests they have in common, even after the signing of a convention, and become, like the regional groups, more or less permanent features of the international scene. After all, the Group of 77 began as a set of states, espousing a common position, at the United Nations Conference on Trade and Development in 1964. But by 1973, and even by 1967, the Group of 77 had a history and could, on occasion, command an ideological loyalty that appeared to ignore interests, (37) as indeed could the regional groups that made it up; or, if a state did not ignore its overall interests, it might deem it prudent to subordinate interests in the specific issues before UNCLOS III to a broader reckoning of the value of remaining in good standing with the group which with it was more permanently associated.

The weight of pre-existing, as opposed to conference-generated, groups, was greater in the First Committee than in the Second. In the former, the main battle-line, as in UNCTAD, was between developed Western states and the Group of 77, with the socialist states, on some issues, creating a further dimension. There was little unity among the Western states, and the UN grouping, Western Europe and others, had no discernible influence as such. The United States, which played the leading role among Western states throughout, was not in it; (38) and some who were in it, like Canada, were land-based producers. There were repeated, and intermittently successful, attempts by members of the European Community to unite behind a common front, on one occasion to claim the right of the EEC, as such, to sign the eventual convention, and on another on the question of what contractors should pay the Authority, where the Community favoured a level of payments lower than anyone else had proposed, including the USA and Japan. Since Evan Luard, the public champion of the "common heritage", was minister in charge of British policy at the time, Britain's adherence to this proposal was some evidence of the pull of community membership.

It was the Group of 77, though, that, in First Committee matters, made the most resolute efforts to present a common front. There is little evidence that the land-based producers within it met as a group, or came into conflict, collectively, with the Group of 77 as a whole, in spite of a clear divergence of interest. (39) The numerical power of the Seventy-Seven, and its possible strategic interest in establishing a reputation for obduracy, have already been discussed.

As a group, though, it does not seem to have ever calculated in these terms. On the contrary, it has often found great difficulty in coming to any decision at all. At Caracas, for instance, the First Committee waited about a week for the Group to agree on its own proposals for the "Basic Conditions of Exploitation" for the sea-bed. It has proved equally difficult for the Group to change a decision once arrived at. In spite of the able leadership of the Peruvian, Alvaro de Soto, the Group rejected, in 1976, the compromise he had negotiated which was embodied in the RSNT. In short, those who wanted to negotiate with the Group, and suggest possible packages which might offer it gains on some issues in return for concessions by it on others, found it difficult to elicit from it what kind of packages might be acceptable.

As the conference proceeded, though, the group developed more competence in such negotiations. Both on the question of the financial terms of contracts, in 1979, and on that of the structure of the Authority, in 1980, representatives (three and six respectively) were able to negotiate on behalf of the group, and in the latter session, the president consulted daily with fifteen of its representatives, who reported daily to the group as a whole, "and the latter did not reverse any of the compromises or commitments it had made" (Commonwealth Secretariat 1982, p.58).

Conference-generated groups were not entirely inactive in the First Committee. The group of Land-locked and Geographically-Disadvantaged States, whose main battleground was in the Second Committee, nevertheless marshalled its forces from time to time in this committee too, notably on the question of representation of their category of states on the Council of the Authority, and special consideration, in the case of land-locked LDCs, in the distribution of revenues. On issues of this kind, the existence of the LLGDS seems clearly to have helped to shape the outcome.

In the Second Committee, where geography was much more central to the negotiations, a quite different pattern of groupings developed, consisting, predominantly, of various groupings of coastal states, on the one hand, and the LLGDS on the other. Even here, there were some important questions on which differences tended to reflect preexisting groupings. The African, and later the Arab, states were prominent in arguing for making the continental shelf extend no further than the economic zone; the Latin American and Caribbean states, who did not all have margins that extended beyond that zone, nevertheless largely tended to favour including such margins in the definition of the shelf. Nevertheless, in most of the Second Committee's work, and specifically that part of it that related directly to the domain or revenues of the Sea-bed Authority, it was the conference-generated groups that came to the fore.

The course of these negotiations will be examined in Chapter 6 below. Suffice it to say here that, while there is no doubt that the Group of 77 was a force to be reckoned with in the First Committee, it was not demonstrably clear that the LLGDS was a force to be reckoned with in the Second, even though that was focus of its chief concerns. It may have been influential, but it was at no point self-evidently so. The reasons for this appeared to be, first, that the LLGDS, as such, possessed no credible threat to act outside the conference; what they could collectively threaten was confined to the conference itself. Secondly, as we have seen, the one threat they did have, to prevent the adoption of a convention by ensuring that the two-thirds majority for it was not forthcoming, was only theoretically within their capacity to execute. Finally, LLGDS members were often, for one reason or another, wary of antagonising powerful coastal state neighbours; those of Eastern Europe, for instance, could not afford differences with the Soviet Union, and there were land-locked states in other continents which were not entirely confident of the security of their access to the sea. Thus any influence of LLGDS on the conference had to come from persuasion rather than power.

The conference-generated groups that were most conspicuously successful, in the Second Committee, were those of various categories of coastal states, such as the Archipelagic Group. Here, as elsewhere, specific interests, even if restricted to a very small number of states, tended to prevail over considerations of benefits that might accrue to the community at large. Even the fact that the community's interest in the international area was to be institutionalised, in the form of a Sea-bed Authority, in the First Committee, did not much influence the way in which the Second Committee considered coastal state claims to extend their jurisdiction at its expense.

Not all groups that have been influential in the conference were straightforward interest-groups. Though a clear, objective, distinction cannot be made between promoters of international order, and potential conspirators against it in the pursuit of sectional interests, some groups seem, on balance, to belong more to the first category that the second, including two cited by Buzan: the disputes settlement group, which contributed to Part IV of the SNT, and the Evensen group, which had operated in the Sea-bed Committee since 1972 (Buzan 1976, p.237). The fact that the president's first attempt at Part IV had such a rough passage in the fourth session suggests that the work of the disputes settlement group was by no means uncontroversial. The Evensen group, too, was seen by some as "oriented almost completely to coastal state interests", (40) and Buzan himself had earlier referred to "the suspicion with which it was viewed" by many of the Group of 77 (Buzan 1976, p.238). Nevertheless, the Evensen Group advanced the work of the conference in two major ways. First, it produced formulations on the substance of certain questions (including the principle that the mineral resources of the continental margin beyond 200 miles should accrue to the coastal state but with some obligations to share its revenues with the authority), which were incorporated in successive texts and thus, effectively, achieved consensus; and secondly, in that its methods of work were adopted, in effect, by the conference as a whole. It was in the Evensen Group that the practice developed of of the chairman submitting a draft of articles dealing with a given issue, listening to comments from all sides, and then submitting a revised draft, which took note, as far as possible, of those comments, which was how the conference itself proceeded from 1975 onwards, and particularly in the fourth session. Engo's attempt, in that session, to achieve consensus though a succession of drafts (called PBEs) was later condemned as "the Committee I Mafia"(Eustis 1977). There was of course a difference in style between Evensen and Engo; but it is not easy to identify a difference between Evensen's negotiating method, and that used by Engo in the spring of 1976.

A further group that might more unequivocally be classified as a promoter of agreement surfaced in the final session, when a number (41) of the less prominent developed Western states attempted to bridge the gap between the demands with which, after its 'Review', the Reagan Administration returned to active negotiations, and the position of the Group of 77 broadly in support of the Draft Convention as it stood. Such chances as their efforts had of success, however, were severely impaired by their harsh reception at the hands of the US Delegation, which one of its leading members was later to suggest might have reflected a "tragic failure of communication" (Ratiner 1982, p.1016).

External Factors

No account of the process by which UNCLOS III attempted to negotiate, among other things, a regime for deep sea-bed mining would be complete without some consideration of what was happening outside the conference. Two developments, in particular, had an important bearing on the negotiating process. One was the dramatic reversal of the trend in the price of oil in the early 1970s, culminating in the fourfold rise of 1973, and the consequent hopes of producers, and fears of consumers, that the prices of other primary products could be made to follow suit. The other was the persistent pressure, on the part of those who had plans to mine the deep sea-bed, for some form of legal acknowledgement of the rights they claimed that would short-circuit the diplomatic process, and give them better, and more secure, terms than they saw themselves getting under any convention that UNCLOS III might be expected to deliver.

The explosive rise in oil prices in 1973 was not an isolated phenomenon. Part, at least, of what made it possible, was the legitimacy which international bodies, going back to the first UNCTAD in 1964, had given to the demands of the developing countries for a transformation of the economic relationships between the world's rich and the world's poor, the producers of raw materials (and other primary commodities), and their consumers. When, therefore, in the early Seventies, the oil-producing countries began to revise their contracts with the oil companies, and assume control of production and prices, military support for those companies in resisting such changes became, almost for the first time in their history, so politically embarrassing as to be out of the question. Thus the price rises, coming on the eve of the first session of UNCLOS III, seemed the fruits of a revolution. They led consuming countries to foresee, with some dread, comparative increases in the prices of nodule minerals, and therefore to dig in their heels against any regulatory powers that could be used to diminish supplies of such minerals by impeding access to the ocean floor. They led developing countries, including even those that thereby suffered heavy increases in their import bills, to welcome the success of the oil exporting countries as an example which might be followed by Third-World exporters generally.

These changes in expectations about the future of the world economy gained added momentum from the two special sessions of the General Assembly the sixth and the seventh of 1974 and 1975. The first, closely preceding the second, and first substantive, session of UNCLOS III at Caracas, was notable for the adoption of the Declaration and Programme of Action on the Establishment of a New International Economic Order, the last four words of which had something of the ring of the "Common Heritage of mankind". It was followed by the adoption of the Charter of Economic Rights and Duties of States, at the 29th regular session of the Assembly in the autumn of the same year, and by a decision by Commonwealth heads of government, meeting at Kingston, Jamaica, in the spring Kingston, Jamaica, meeting at Commonwealth Heads of Government, 1975 of 1975, to "take immediate steps towards the creation of a rational and equitable new international economic order", and appoint a group of experts to draw up a "comprehensive and inter-related programme of practical measures directed at closing the gap between the rich and the poor countries". (42)

A further special session of the General Assembly, the Seventh, immediately preceding its regular session of that year, considered this Group's report, to which representatives of developed and developing countries alike had contributed. In contrast to the starkly confrontational atmosphere of the 1974 special session, it was characterised by a mood of compromise.

These institutional activities were important for two reasons. First, they helped to set the stage for the sea-bed negotiations at UNCLOS III, which were seen quite unequivocally by the Group of 77 as a step towards the New International Economic Order. The international area of the sea-bed belonged to mankind as a whole; it was to be exploited in such a way as to reflect the general principles by which the world economy was to be reformed, principles which in some cases, at any rate, had secured apparent consensus among rich and poor countries alike, and failing that had certainly been overwhelmingly endorsed by global political bodies. Though the industrial countries did not, for the most part, accept that the regime for the sea-bed should be conceived of as an element in the New International Economic Order, or shaped by its tenets, it was clearly impossible for UNCLOS III to adopt any convention whose sea-bed provisions could not be legitimised by reference to it. (43)

The second reason why these events were influential on UNCLOS III was their specific implications for commodity arrangements. The secretary-general of the United Nations Conference on Trade and Development had already recommended the establishment of an integrated programme of commodity arrangements financed by a 'common fund'. This was strongly endorsed by the Commonwealth Group of Experts and referred to, though in less committed terms, in the resolution adopted at the Assembly's seventh special session. At the fourth UNCTAD, in Nairobi in 1976, the Integrated Programme and the Common Fund were the main theme of debate, and though no agreement was reached on the details UNCTAD committed itself to the principle of such a fund and promoted a series of negotiations, which were to bear fruit, eventually, though on very much more modest scale than had been originally proposed, at its third negotiating session in March 1979.

The world, and more particularly the Group of 77, was thus committed to some form of regulation of world commodity markets. This could hardly be reconciled with the principle of free access to the minerals of the sea-bed on which the developed states insisted. Thus the main protagonists, in the debate over the sea-bed regime in UNCLOS III, could hardly fail to read into that issue intimations of more general considerations on which they were fundamentally at variance.

Discussion of the negotiating process by which the attempt was made to transform Pardo's item of 1967 into flesh and blood, or at any rate, a regime and machinery, a convention and a sea-bed authority, naturally prompts the question: could it have been done better? Are there points at which decisions taken lightly and in ignorance of their implications impelled the negotiations along roads they would otherwise have not taken, or confronted them with obstacles they might otherwise have avoided?

Given that the conference lasted far longer than it was expected to (when it opened, 1975 was seen as the year of decision!) and even then did not manage to produce a generally-agreed convention, it would seem wise to reflect on why its results so disappointed expectations.

The starting-point for any explanation of the conference's inability to achieve the goal it had set itself must lie in the discontinuity in United States policy. What was acceptable to the US Delegation in the summer of 1980 was not acceptable to the US, following the change in administration, in the spring of 1981 and thereafter. It is an inherent and recurrent difficulty in the negotiation and maintenance of international agreements that a position accepted by a government to-day may be repudiated by its successor tomorrow. This extends even to the point of violating treaties to which a state is a party. The Reagan Administration may have wrecked the conference; it may indeed have cast a shadow over future attempts at multilateral negotiations; but it did not, at UNCLOS III, put the USA in breach of international law since the USA had not got to the point of committing itself, in any legal sense, to the text. Had it done so, of course, full legal commitment would have required ratification by the Senate. Since only one-third of American senators are elected at any one time, the United States, in the treaty-making provisions of its constitution, is better protected than most states against the possibility of one administration giving legal undertakings that its successor cannot accept, though correspondingly more likely to have to negotiate through administrations which prove, like Woodrow Wilson's in respect of the League of Nations after Versailles, to be unable to implement the bargains they have struck.

The latter aspect was particularly frustrating in a protracted conference like UNCLOS III, where renegotiation of a once-agreed text involves so many other parties. Nevertheless, the possibility that American administrations will take a different position from their predecessors, or not be supported by the Senate as concurrently constituted, is a fact about American politics to which the conference did not give due weight. A not dissimilar fact of the diplomatic world, which also caused the conference some delay, is the habit of goverments - especially those who have newly come to power - to recall diplomats used by their predecessors, even if those diplomats happen to hold office - in Amerasinghe's case, that of president of the conference. Here delay was compounded by ambiguity. For the Latin Americans, Amerasinghe could clearly not remain president in 1978, when he ceased to be a delegate. For the Africans and Asians, he was seen as an individual in whose presidential qualities they had confidence, and who should therefore see the conference out. The conference would have saved itself nearly 3% of its sessional time, if it had clearly established a rule favouring one or other of these positions before the occasion arose.

Had UNCLOS III concluded its business more rapidly, these effects of changes of government among its membership might have been avoided. There might have been a time when a US administration could have persuaded the conference to adopt a treaty that it could also have persuaded the Senate of the day to ratify, and if so we must ask what prevented it; but we must also ask whether, even in that same case, a less sympathetic successor in the White House could still have managed to sabotage a treaty which was seen as inimical to US interests, even if the US was now a party to it. If the answer is yes, then what was needed was the negotiation of a treaty sufficiently advantageous to US interests, or sufficiently flexible in the obligations it imposed on the US if it became a party (that is, sufficiently "reversible" as to avoid being characterised as hostile), to such interests by any subsequent administration that was at all foreseeable; in other words, either a text which, as compared with the convention, was much more steered in favour of the USA; or one which imposed much less drastic commitments on _all_ states. It is difficult to see the Group of 77 accepting the former; and accepting the latter would have meant aiming at a rather different target.

Much of the negotiations we are concerned with in this book can, by a drastic simplification, be reduced to bilateral negotiations between the United States and the Group of 77. The inability of the USA, over the years, to sustain a coherent position in these negotiations was to some extent mirrored by the difficulties encountered by the Group of 77, which have already been described. Again, the remedy does not wholly, or even mainly, lie in the Group's electing a sufficiently accommodating and realistic negotiatior, or negotiating team, to strike a good bargain when it is offered, although there is clearly room for improvement in the Group's negotiating arrangements in the regard. Success required that what the leaders of the Group agree to should be accepted to its constituents, not just momentarily, but when, as a result of one government replacing another, or other changes in circumstances, they take a fresh look at their law of the sea policies. Again, a convention that might meet this requirement might need to be more modest in its commitments than that at which UNCLOS III aimed.

In any case, the simplification involved in treating the process of negotiating a sea-bed authority as a bilateral interchange between the USA and the Group of 77 is Procrustean. In form, at least, it was a multilateral negotiation, and that form was one to which most, if not all, delegates jealously stuck. What emerged had to emerge from a process in which all had the opportunity to participate. From this flowed the delicate role the president and the three main committee chairmen were to play, and the recurrent oscillation between public and private, and inclusive and selective, meetings. The adoption of President Amerasinghe's 1975 suggestion that negotiations should henceforth proceed on the basis of chairmen's texts put more weight on the calibre of those chairmen than might have been reckoned with when they were appointed. Had someone other than Paul Engo chaired the First Committee, much faster progress might have been made after the 1975 and 1977 sessions. If the idea of chairmen's texts is to be regarded as a constructive innovation in multilateral diplomacy, applicable to many other conferences besides UNCLOS III, it needs to be accompanied by more careful selection of committee chairmen and provision for the easier dismissal of those that, by interfering with what might otherwise be successful mediation attempts, become liabilities rather than assets.

The tension between meetings of all states and meetings of a select few was even more difficult to resolve in UNCLOS III. It was one thing for members of a regional group to agree that one state should speak on their behalf at a meeting at which they were also present. It was much more difficult for them to agree that it should represent them, at a meeting, concerned with substantive questions, at which they were not present. In the end, if the First Committee's Group of 21 can be taken as a test-case, the "arena" method, tried without much success at the fifth session in 1976, proved the most productive. Each region was, somehow, assigned representatives, and alternates, to whom the discussion was normally confined, but all other states could attend, and might, by taking a conspicuous initiative, also on occasion contribute. This arrangement, implemented at closed meetings from which non-governmental representatives were excluded, paralleled the division in open meetings between delegates and observers, including non-governmental observers.

It could also be claimed that the difficulties of negotiating a sea-bed authority were compounded by the awkwardness of having to combine such negotiations with a comprehensive rewriting of the rest of the law of the sea. Against this, it was claimed that the creation of a gigantic "package deal" with something in it for everybody was more likely to win universal support than a conference confined to implementing the "common heritage" concept. As we shall see, it was originally the developing countries, and especially the Latin Americans, who linked the coastal state revolution with the proposal to establish a common heritage of mankind. The maritime states – that is those that tended to use the sea off other states' coasts – went through three phases in their response to this. First, they opposed the wider conference because they feared it would lead to more extensive coastal state jurisdiction; then they accepted the idea of coastal state jurisdiction, subject to restrictions on interference with navigation, which were written into successive texts. They were then sufficiently in favour of the wider treaty, because of the provisions of Parts II, III and IV, to be prepared to accept, in Part XI, provisions for creating a seabed authority with substantial jurisdiction over sea-bed mining as an inducement to all states to become parties to the treaty and abide by it. This was the most hopeful period for the chances of a comprehensive treaty, in which for instance, the US Defence Department was a strong advocate of an accommodating line. In the third phase, it began to be argued that the maritime powers did not need to have a treaty in force in order to be secure in the rights of navigation it embodied, on the grounds that these had now become part of customary law. Thus inducements to coastal states to join the treaty were unnecessary. In particular, under Reagan, the Defence Department became an opponent of the treaty (Ratiner 1982, p.1011).

These changes in the evaluation of the package were probably, on balance, disadvantageous to the prospects of agreement; but the complications such a package posed cannot be said to have delayed the adoption of a convention, since the rest of the text was almost completely agreed while the sea-bed negotiations remained recalcitrant.

Deadlines, or their absence, have also been blamed for the disappointments of UNCLOS III. Ed Miles, as we have seen, listed the fact that the General Assembly resolution convening the conference did not impose a deadline for final decision as one of the eight factors accounting for the extremely slow pace of the Caracas session.

"[it] appeared to reinforce the reluctance to compromise on positions that are strongly held because country representative generally do not wish to yield until the very last moment" (Miles 1975, p.45).

This is no doubt true; but the absence of a deadline may also have permitted compromises to have been worked out which could not have been reached within one. The 1980 Draft Convention, on which there was, at the time, something very near consensus, could hardly have emerged from UNCLOS III had the latter been required to end its deliberations by, say, 1975, which would surely have led to a text carried, on many points, by the sheer numbers of the Group of 77. In 1981, however, a deadline was imposed by the ambiguous decision simultaneously to formalise the text and allow not more than three weeks of the 1982 session for further informal negotiations. The conference, by consensus, committed itself to making that the final session. This may have been, with hindsight, excessively rigid. Certainly it left little time for mediating between the extensive American amendments, now supported by most of the other Western states, and the Group of 77 and other defenders of the 1980 text. But as Leigh Ratiner, now once more charged with implementing US policy on ocean mining questions, was subsequently to make clear, such time as there was was not used, by the United States, to best effect:

"Our strong stance on every issue... persuaded the
bulk of Conference participants that the US appetite
was too great - no improvements were likely to
satisfy us that could also be swallowed by the Third
World" (Ratiner 1982, p.1013),

and again,

"After the United States apparently rejected the
good samaritan papers as a basis for negotiations,
the remaining weeks of negotiation were carried out
... in a desultory and pessimistic atmosphere, even
though time permitted serious negotiation of the
main issues of concern of the United States"
(p.1016).

Eventually, though, after the US Delegation had secured a
change in its instructions, intense negotiations were
resumed and appeared to come close to success (Ratiner 1982,
p.1014).

A final major question mark is over the degree of detail
in which the regime was negotiated. The Group of 77 wanted
a powerful Authority and one in which numerical strength
would preponderate; the West, and, in particular, the
United States, wanted a body of limited powers and one in
which their representation reflected their economic and
technological strength. Logically, two compromises are
possible: a body with wide discretion in which the West
would have collectively enjoyed a dominant role; or one
whose decisions would be largely controlled by the numerical
majority of developing statues, but was limited in its
powers. The West chose to emphasise the second kind of
compromise, which necessitated detailed negotiation of the
financial obligation of contractors and the conditions under
which the Authority could refuse an application for a site.
In the structure of the authority it sought only a blocking
minority in the Council. It is just possible that, had the
other compromise been pursued, much of this detailed
negotiation would have been unnecessary. If the Authority
were so structured that the West could have been confident
that its decisions would, on the whole, have regard to
interests as ocean miners and consumers, it could perhaps
have been given more discretion.

Such a compromise though, might have been unacceptable to
either side. The United States Senate, if not its
Administration, might have objected to regulatory powers
being given even to a body dominated by Western states. The
Group of 77, too, was, as we have seen, equally committed to
the 'democratisation'of internal economic institutions on
the principle of one state one vote. An international
authority which could conceivably have appealed to the West,
as a body through which, because of its structure, they
could hope to promote the policies they thought necessary,
would almost certainly have grossly violated this principle.

The analysis offered in this chapter suggests that the failure of UNCLOS III to achieve consensus in 1982 was probably not preordained, though the task it had set itself was an exceedingly formidable one. Accidental factors, impediments which might be avoided in another multilateral negotiation of this kind, could account for the fact that the convention was not adopted by consensus, early in the Carter Administration; had it been, the Preparatory Commission set up under it might quite possibly have produced 'rules, regulations and procedures' sufficiently acceptable to win ratification from all sides, including the USA., if a state's failure to ratify was to mean casting some doubt on the legal validity of its nationals' claims to exploit sites they had already spent some years exploring. These are fairly tenuous "if onlys", but they are not wholly fanciful. What seems undeniable is that the implementation of the concept of the common heritage, which proved the most intractable element of the agenda of UNCLOS III, would in any case have been a protracted process. It might have been negotiated in seven or eight substantive sessions instead of ten;(44) but nothing which might hope to carry with it global legitimacy could have been accomplished in a mere one or two.

The monumental stamina on the part of delegates and governments, at UNCLOS III is not, in itself, unusual. Disarmament negotiations, since 1945, readily demonstrate how endlessly states and their representatives can sustain diplomatic efforts with pathetically meagre results. Those at UNCLOS III were, however, more dynamic than most of those on disarmament over the last few decades. There was movement, and, for the most part, a sense of cumulative, if gradual, achievement. The commitment to making a reality of the common heritage at UNCLOS III (though not, perhaps, the prime objective of most delegations) was more than a ritual acceptance of a socially respectable target; it represented a creditable willingness to imagine, and fashion, a world within which national interests were made compatible with some conception, albeit limited, of the common good.

NOTES

1. The General Committee, in the case of the UN General Assembly.

2. The Economic and Social Council, for instance, had already in March 1966 asked for such a study, which was published, as E/4449, in February 1968 (Buzan 1976, p.66, and n. 1 on p.88).

3. Buzan, (p.71) lists one "major report" and five "shorter reports" produced by the Secretariat in 1968 alone for the Ad Hoc Sea-Bed Committee. This spate of studies continued throughout the life of the SBC and indeed UNCLOS III itself.

4. An example, drawn from another context, would be Trygve Lie's controversial memorandum on the question of Chinese representation circulated to the members of the Security Council other than Nationalist China in 1950 (Ogley 1964, p.590).

5. Excluding those of the Soviet Union, in Central Asia, which have been with some logic, so described, but which have never identified as such by the anti-colonial movement in the United Nations.

6. Hjertonsson 1973, p.42, relying on a "trustworthy Latin American source". Leigh Ratiner also traces UNCLOS III back to consultations between these two powers in 1966 (Ratiner 1982, p.1007).

7. Its full title, originally, was "The Ad Hoc Committee to study the peaceful uses of the sea-bed and the ocean floor beyond the limits of national jurisdiction". The words "Ad Hoc", implying that it would last only long enough to report to the next session of the General Assembly, were dropped in 1968. The words "to study" were also omitted and replaced by "on".

8. 43, in practice. A seat was reserved for the Eastern European group, but the group declined to nominate its sole remaining member, Albania, to it.

9. Resolution 2340 (XXII) of 18 December 1967, operative paragraph 2 (c).

10. Resolution 2467 (XXIII) of 21 December 1968, operative paras, 2 (a), 2 (b) and 4 (b).

11. By 65 to 12 with 30 abstentions.

12. Resolution 2750 C, para 2.

13. Ibid., para 6.

14. Oda p.105. Thus even the Sea-bed Committee, itself a selective body, was manifesting a dilemma which was to recur repeatedly in subsequent negotiations at UNCLOS III: a body in which all states attending the conference participate was impossibly unwieldy, a body confined to a supposedly central few aroused (on the part of those excluded) resentment and a suspicion of the accommodation reached within it.

15. This might have happened anyway. But if somehow, a conference had been convened on sea-bed mining alone, or if the Sea-bed Committee had give unequivocal priority to the settling the boundary of the international area, many coastal states, particularly in Latin America, might have been brought into more direct opposition with those championing the "common heritage of mankind", and the coastal states themselves might well have been split.

16. In 1971 and 1974.

17. The then leader of the British delegation, Sir Roger Jackling, even declared that the debt the conference owed its president, for fostering this compromise, was "unequalled" at any previous conference.

18. A/Conf. 62/30/Rev 1. Rules of Procedure, Rules 37 and 38.

19. Finishing 15 July 1974. See UN/Press Release SEA/C/49, 15 July 1974.

20. On this occasion, partly because delegates were asked to confine themselves to not more than ten minutes each, delegations were explicitly authorised to present written statements on the current stage of the conferences work. Eight (Paraguay, Romania, Mongolia, Canada, Argentina, Peru, Bahrain and Guatemala) did so. This variant of public debate was to be used again.

21. It was later made public in the course of Congressional hearings in the United States.

22. The title emphasised the subordination of this group in theory, and as it proved in practice, to the First Committee's official "chairman", Engo.

23. By Algeria and Peru at the 34th meeting of the First Committee, 9 September, 1976; see UNCLOS III Official Records, Volume VI, pp.69-70. The ominous implications of this ostensibly mild demurral may however, have been apparent only with hindsight.

24. It was reported that the delegation had not prepared a position on the structure of the Authority which, had the RSNT been accepted, would have been the chief raison d'être of the session.

25. A/Conf. 62/62 of 13 April 1978, paras 3 and 4, in UNCLOS III, Official Records, Volume X, pp.6-7. In his report to the plenary of 19 May 1978 (A/Conf. 62/RCNG/1), Engo made clear that he had exercised his discretion in this respect.

26. A/Conf. 62/62 of 13 April 1978, para 10. On the face of it, this could be seen as a strengthening of the hands of the committee chairmen, and especially of Engo, since it entrenched the ICNT, which bore their stamp, against contentious change. In fact, it enabled subsequent revisions to be made by consensus, and entrenched them, against arbitrary alterations by the chairman. As late as the first part of the Ninth Session, however, the US Delegation was to complain of just such an alteration; that is, that a contradiction of these criteria without enjoying the necessary support. (see US Delegation Report, Ninth

Session, February 27th to April 4, p.8 & 28).

27. See above. The position of Engo as First Committee Chairman was, in form at any rate, somewhat modified by giving the chairmen of Negotiating Groups 1 and 2, and of a Group of Legal Experts set up to look at the legal problems rising out of Part XI, the status of "co-ordinators of the Group of 21". Engo himself was chairman of Negotiating Group 3.

28. WG.21/1.

29. A/Conf. 62/WP10/Rev.2 of 11 April 1980, and A/Conf. 62/WP10/Rev.3 of 22 September 1980.

30. April 2 and 3.

31. A/Conf. 62/L36, in UNCLOS III, Official Records, Vol.XI, p.96.

32. The Preparatory Commission, and PIP, directly relate to the theme of this book and are treated more fully in Chapter 11.

33. The proposal, by Australia, Canada, Chile, Iceland, India, Indonesia, Mauritius, Mexico and New Zealand (A/Conf. 62/L4) was dropped as such after the President's ruling to permit its presentation to plenary had been upheld by only 50 to 38.

34. Official Text: interview given by Elliot Richardson to Hugh Muir, Wednesday January 11th 1978 (American Embassy 1978, pp.1-2).

35. "If these is no objection I shall take it the Conference approves", he would say, having made a procedural proposal; and then, after only the briefest pause, bring down his gavel and say "I hear no objection, the Conference has approved".

36. For a more extended discussion of the role of these groups, see Buzan 1980, and Miles 1976.

37. See, for instance, Algeria's contribution at Caracas to the debate in the Second Committee on the Economic Zone, cited in Miles 1975, p.42.

38. Though it has been included in that group for the purpose of Table 4.1 in the previous chapter.

39. In that developing states in general sought to maximise their participation in, and revenues from, sea-bed mining, while land-based producers might have done best had there been no sea-bed mining at all. See above, Chapter 4.

40. Miles 1976, p.176. It was certainly not especially

sympathetic to, or representative of, land-locked states, of which there were at most two, and sometimes only one, out of a membership of 25 to 30.

41. Variously given as 10, 11 and 12.

42. The 'interim report' of this Group, entitled "Towards a New International Economic Order", was submitted on 23 July 1975 and published by the Commonwealth Secretariat in the next month.

43. The controversial character of this requirement can be seen by examining the provision of the 1974 'Declaration and Action Programme' that "... the developing world has become a powerful factor that makes its influence felt in all fields of international activity. These irreversible changes in the relationship of forces in the world necessitates the active, full, and equal participation of the developing countries in the formulation and application of all decisions that concern the international community" (Spero 1977, p.179). This has clear implications for the structure of the international authority quite at variance with what the ocean mining states saw as essential features of any body which could command their confidence. See ch.9, below. Nevertheless, as far as I have argued elsewhere (Ogley 1981) the negotiations at UNCLOS III, and the Draft Convention of 1980 that emerged from then, do not represent anything like the straight application of the NIEO to the Law of the Sea. Even the Group of 77's demands at UNCLOS III had important elements unique to that conference.

44. The first, in 1973, being wholly procedural.

PART II
The Issues

6 The Geographical Boundary of the Common Heritage

To set up an international sea-bed authority, answers have to be given to three exceedingly vexed questions: within what geographical limits shall it exercise jurisdiction, or collect revenues? Within these limits, who shall be permitted to exploit the sea-bed, in what circumstances, and subject to what regulation and oversight? and how shall the sea-bed authority be set up, that is what organs, of what composition, and by what majorities, shall decide what questions, and to what appeal mechanism, if any shall their decisions be subject? The history of the international community's attempts to answer the first of these questions will be traced in this chapter; that of the second, in chapters 7 and 8; and that of the third, in chapters 9 and 10. In each case, the historical account of the handling of each question will be preceded by some consideration of the question in its own right.

There would be little point in setting up a sea-bed authority without agreeing on clear rules by which the boundary of its domain, that is to say the "International Area", was to be defined, and the limits to coastal state jurisdiction set. On the other hand, questions of limits can be, and between 1945 and 1967 were, discussed without any implication that what lay outside them should be subject to the control or supervision of an international organisation; and even after 1967, for those who claimed, as did the companies described in chapter 4, that ocean mining beyond the limits of national jurisdiction is one of the "freedoms of the seas", the "limits" question was quite separate from that of "the regime and machinery". This separation was reinforced by the division of functions, in the Sea-bed Committee and UNCLOS III between the First and Second Committees, so that limits could be discussed in the Second Committee without any reference to what was happening in the First.

Those coastal states that stood to gain most from the broad definition of limits (1) were not therefore directly confronted with the effects of their claims on the common heritage of mankind. Others, that might have hoped to gain more as participants in, and financial beneficiaries of, collective exploitation of the larger international area that would have resulted from denying those claims, than as fellow-claimants to more extensive national jurisdiction, have preferred, in effect, to take a chance. An additional tract of unshared sea-bed for themselves might or might not contain minerals of value; a narrower definition of the boundary, by which that tract, and all others like it, became part of the international area, was much more certain to add something to mankind's collective wealth. The fact that, on limits, states generally seem to want to take chances of this kind suggests that such negotiations are not simply about wealth, but have a quasi-territorial nature, different in kind from all the other issues examined in this book. They seem to rearouse some of the same passions that break out in territorial disputes on land.

THE CRITERIA FOR LIMITS; SOME THEORETICAL CONSIDERATIONS

There are, in principle, four kinds of boundary between the international area and that subject to national jurisdiction: first, a limit based solely on characteristics of the sea floor itself - depth, gradient or type of rock, for example; secondly, a limit based solely on distance from the coast (or from baselines); thirdly, a combination of sea floor characteristics and distance, but giving a single boundary line between national and international jurisdiction throughout; and fourthly, a "blurred" limit, that is a set of two or more boundaries creating a zone or zones in which national and international jurisdiction would in some way be combined. The first of these possibilities has never, in practice, been considered. All claims based on characteristics of the sea floor have related only to what lies outside the territorial sea, whose extent has always been determined by distance, and within which coastal state jurisdiction over the sea-bed has been undisputed. (2) The question thus becomes to what extent, if at all, and, if so, how, should sea floor characteristics be combined with those of distance in establishing the boundary. (3)

There are two main arguments in favour of bringing in such characteristics, though they do not point in exactly the same direction. The first is based on a claim to precision. Depth, in practice, is more precise than distance, because distance, in the law of the sea, means distance from baselines, and coastal states have traditionally advanced a variety of specious justifications for defining baselines in idiosyncratic ways. Depth, though, as we have seen, (4) cannot reliably be used to distinguish the shelf from the slope, the slope from the rise, or the rise from the deep ocean floor, since these submarine features begin and end at different depths in different parts of the world. Moreover, there is usually no clear point at which the slope or rise ends. The only one of these transitions that would have made a precise boundary would have been the edge of the shelf, measured in terms of change of gradient.

The second argument for emphasising sea-floor characteristics rather than distance is that, as we have also seen, they seem to correspond neatly with the way in which the wealth of the sea-bed is distributed. Hydrocarbons, it is thought, are likely to be found only in the continental margin; commercially-attractive manganese nodules, only in the deep ocean floor far from land. This argument assumes that it is preferable that the Sea-bed Authority should deal in only one kind of resource, and (insofar as it was an argument against the creation of a 200-mile economic zone) that it should have a monopoly of it. It thus cuts no ice with those who, like Dr. Pardo, sought to make oil and gas part of the "common heritage of mankind". Moreover, because there is no precise measurement of the edge of the margin, particularly if it is defined so as to include the continental rise, precision can be attained only at the cost of arbitrariness either by using a depth limit (of 2,500, 3,000 or 4,000 metres), or, less satisfactorily, by reference to depth of sediment, distance from the foot of the slope, or some combination of the two.

The case for relying on a distance criterion is partly based on the difficulty of finding an adequate alternative, and partly on its inherent merits, which are chiefly two. First, whatever the configuration of the sea-bed off their coast, coastal states are interested in activities near their shores. A mining operation of a given kind a given distance from shore will affect the interests of coastal states equally, whether it is in waters of 100, 1,000 or 4,000 metres. It is distance rather than depth that determines how likely it is to pollute its beaches, or interfere with navigation or fishing. Conversely, an operation in shallow water many hundreds of miles off land might well affect the interests of other users of the sea more that those of the state of whose shelf it might be said to form part.

It can therefore be argued - and this is the other merit of a distance criterion - that it would make sense to have a single limit to coastal state jurisdiction for all purposes - or at least for all resource-connected purposes, including the taking of fish, any edible denizens of the shelf, and the minerals on and under the sea-bed, and, if any were worth exploiting, in the sea itself. A single boundary line of this kind would be simpler for all concerned: the national and international authorities, and those subject to their jurisdiction. It would also avoid the possibility that the activities of a state, or its nationals, in what remained as high seas, might be inconvenienced by mining installations which it had no say in regulating because they formed part of some other state's shelf or margin. (5)

The main disadvantage of a distance line has already been mentioned in passing. It is that of imprecision, derived from the absence of clear, generally-accepted rules governing what the distance should be measured from, a legal laxity which has, of course, equally plagued orderly determination of the outer limits of the territorial sea.

Much of the anomaly arises out of the baselines used, which in general are supposed to be drawn along the low water line along the coast. In fact, many exceptions are made to this simple and honourable principle (Pardo and Borgese 1976, pp.22-28). States often claim the right to draw straight baselines, particularly, though not necessarily only, where the coast is deeply indented, or where there is a fringe of islands in the immediate vicinity of the coast. (6) The circumstances in which such baselines can be drawn, and the questions of whether any limit is set to their maximum length, or whether they must be drawn from land points, or can be drawn from low-tide elevations if installations permanently above sea-level (eg. light-houses) have been built on them, or from any low-tide elevations, or even from "appropriate points" in the sea some distance from the coast, all make a substantial difference to where the boundary between national and international jurisdiction will lie. By the end of 1972, according to one authority (Prescott 1975, p.78), forty-seven countries had proclaimed straight baselines along part or all of their coast, and for at least six of these countries the longest "leg" exceeded one hundred miles, and for three of them was at some point more than fifty miles from the nearest coast. (7) This boundary can also be affected by claims to "historic bays", which if conceded, would make them internal waters. Prescott (1975, p.98) also lists sixty-three bays in twenty countries "generally considered to have been claimed on historic grounds" (11). They include Hudson Bay (area 472,000 square miles) and several Arctic seas north of the Soviet Union (the Chukchi, the East Siberian, the Kara and the Lapteu).

The supposed precision of distance-based limits is further eroded by the claims of two further categories of states. Archipelagic states claim the right to draw straight baselines connecting the "outermost drying reefs" of their archipelagoes, (8) and the possessors of isolated "islands", habitable or not, claim zones in respect of them as well as in respect of their main territory. Thus what counts as an archipelago, or a drying reef, or an island, has to be decided, before the implications for the extent of the international area of any given distance criterion can be established. (9)

The fact that a simple distance criterion does not lead to a neat division of the ocean by type of resources can also be used as an argument against it. If there were few, if any, prime nodule-mining sites under coastal state jurisdiction, the Authority's bargaining power in the use of its discretion, if it had any, would be that much stronger. If there were little, if any, exploitable oil or gas in the international area, the Authority would lose potential revenues, but it would save the cost of developing a capacity either to embark on drilling operations, or to regulate them, and there would be no danger that a pool of oil straddling the boundary would be simultaneously, and wastefully, tapped by both the Authority and the coastal state in question. On the other hand, a share in the exploitation of oil and gas would give the Authority expertise in that field, which it could share with developing coastal states; and conversely even with nodule mining, competition from coastal states might help to promote its efficiency. Moreover, while there could be no fields of oil or gas straddling the boundary of the international area, if the latter contained no oil or gas, there might well, in consequence, be even more fields that straddled the boundary between adjacent coastal states, so the problem itself would not be diminished.

It can also be said in criticism of a distance criterion, though less convincingly, that it ignores the supposedly self-evident right of a coastal state to the natural prolongation of its land area, however far that might extend. This doctrine, favoured by international lawyers, and upheld, with some qualifications, by the International Court of Justice in the North Sea Continental Shelf Cases of 1969, has some plausibility when, as in that case, its application is limited to the geographical shelf, whose edge constitutes the most clearly identifiable boundary between submerged continent and ocean depths; but even that plausibility lacks practical force as applied to a shelf several hundred miles from land. Viewed objectively, it has little to commend it in such a case, still less when it is used to justify a claim to the margin. It then becomes scarcely more than an instrument for the satisfaction of coastal state greed. (10)

Although, as has been shown, distance limits are necessarily imprecise, some indication of the effect of each of two such limits (40 miles and 200 miles) and of two limits based on depth (200 metres and 3,000 metres) is given by a report produced by the Secretary-General, on the request of the General Assembly, in 1973.

He also estimated how the mineral resources of the sea-bed would be divided between the two jurisdictions by each of the four limits. By all four limits most of what were then seen as the total ultimate resources of hydrocarbons would be assigned to coastal states, 59% by the 40 mile limit, 68% by 200 metres, 93% by 3,000 metres, and 87% by 200 miles.

Boundaries of 200 miles or 3,000 metres would assign all "proved reserves and immediate prospects" to coastal states; 200 metres "almost all", and 40 miles 90%. All known mine-grade deposits of manganese nodules would fall into the international area if the boundary were 40 miles, 200 metres, or 3,000 metres; if it were 200 miles there would be some mine-sites under national jurisdiction adjacent to volcanic islands in the North and South Pacific, but most deposits would still fall to the international area.

Table 6.1

Size of International Area and of the Area Accruing to Coastal States under a Variety of Limits to Coastal State Jurisdiction.

Limit	Size of International Area (sq. km.)	Size of Area Accruing To Coastal States (sq. km.)
40 miles	346,870,000	15,660,000
200 metres	340,360,000	21,900,000
3,000 metres	318,150,000	45,420,000
200 miles	288,040,000	77,080,000

Source
(UN General Assembly 1973 "Economic Significance", Table 5 p.39)

There is, then, little to choose between lines based simply on distance and those based on easily measurable characteristics of the sea-bed, as theoretically possible boundaries between the international area and what should accrue to coastal states. If anything, the balance of the argument favours the former, in spite of their imprecision.

THE DEBATE OVER LIMITS

Before UNCLOS III

But it must not be supposed that the debate over limits, before and at UNCLOS III, has been, in essence, a debate over the principles on which limits should be based. What has happened, in brief, is that, first, depth criteria were used, by those that it suited, to appropriate the shallower part of the sea-bed, without encountering any very strong objection; then distance was used, by others, and ultimately accepted, to encompass an even greater annexation of the ocean floor in the form of a 200-mile economic zone; and then those that could claim anything that could conceivably be called continental margin extending beyond this zone, did so, and won most, if not all, of what they asked for, at the cost of having to share with the sea-bed authority a fairly small part of any revenues derived from exploiting it. The general effect has thus been one of massive expansion of the area under coastal state jurisdiction, and so far it has only been under coastal state jurisdiction that the sea-bed has been commercially exploited at all. The prospect from 1967 onwards of managing the area beyond jurisdiction on behalf of mankind as a whole, far from inhibiting coastal state expansion, has seemed rather to accelerate it.

In spite of some scattered instances of previous claims or agreements relating to the sea-bed, notably the Gulf of Paria Treaty between Venezuela and the United Kingdom (for Trinidad) in 1942, the trend towards claims based on sea-bed characteristics, in this case depth, and extending beyond territorial waters (or providing the occasion for the extension of such waters) may be said to have begun with the Truman Proclamation of September 25 1945. This asserted the exclusive jurisdiction and control of the United States over:

"the natural resources of the subsoil and sea-bed of the continental shelf beneath the high seas but contiguous to the coast of the United States."

An accompanying memorandum interpreted the "continental shelf" as "submerged land, contiguous to the continent ... covered by no more than 100 fathoms (600 feet) of water". This was not to affect:

"the character as high seas of the waters above the continental shelf and the right to their free and unimpeded navigation".

The principle of a depth limit to coastal state claims to sea-bed resources, beyond what was traditionally recognised as territorial waters, was widely supported when the United Nations, through its International Law Commission, came to take up the question in 1950. In the Continental Shelf Convention that emerged from the United Nations' first Law of the Sea Conference (UNCLOS I) in 1958, however, a limit of this kind was combined with what came to be known as the "exploitability criterion". The first article of the convention defined the continental shelf, to whose resources the coastal state was to enjoy exclusive title, as:

"the sea-bed and subsoil of the submarine areas adjacent to the coast ... to a depth of 200 metres or beyond that limit, to where the depth of the superjacent waters admits of the exploitation of the said areas".

104

Thus there was to be a depth limit, only slightly more generous to coastal states than the Truman Proclamation, and not widely at variance, for the most part, with the edge of the shelf, that is, the point of sharpest change of gradient; yet the force of this limit was apparently negated by this final phrase, which appeared to entail that it did not apply to any part of the sea-bed that subsequently became exploitable.

The delegates at UNCLOS I, and the members of the International Law Commission that prepared the ground for it, seem to have played with the concept of exploitability with the unthinking innocence with which children might play with a cobra, totally unaware of the fatal consequences awaiting them as soon as it raised its head. If everything that was exploitable belonged to the coastal state, what limits were there to be? The words of Article I of the 1958 Convention seem to warrant the gloomy observation of the Japanese jurist, Shigeru Oda, that:

> "all the submarine areas of the world have been
> theoretically divided among the coastal states, at
> the deepest trenches" (Oda 1968, p.9).

Yet the debates at Geneva in 1958 suggest that this was far from being the intention of delegates. A South Korean proposal to make exploitability the sole criterion was overwhelmingly defeated (42 to 13 with 13 abstentions), and several proposals for a deeper limit (550 meters) were also rejected (Buzan 1976, p.39).

It can be inferred that, in the main, delegates voted for the exploitability criterion only because they saw it as being heavily qualified by such words as "shelf" and "adjacent", as well as by the 200-metre figure. According to Sreenivasa Rao, the criterion was first put forward by J.L. Brierly and Manley O. Hudson in the International Law Commission in 1950, (11) and Israel, in 1952, was the first state to base a claim on it (Sreenivasa Rao 1975, p.49). There was considerable opposition to it in the Commission, although some of those who wanted a fixed limit conceded that the latter was not final, and might well need to be revised as technical capacities grew. A decisive feature of the debate was the Resolution of Ciudad Trujillo in March 1956, at which twenty American states, with many individual reservations, endorsed the principle of exploitability. This was enough to turn the scales in the International Law Commission that same year, but only after the Commission's Cuban Chairman, F.V. Garcia-Amador, had, according to the official records, rebutted the objection that such a criterion would "tend to abolish the domain of the high seas" with the argument that "the words 'adjacent to the coastal state' placed a very clear limitation on the submarine areas covered by the article. The adjacent areas ended at the point where the slope down to the ocean bed began, which was not more that 25 miles from the

105

<u>coast</u>" (Sreenivasa Rao 1975, pp.52-3 and n.33).(12)

Even if the Continental Shelf Convention of 1958 had been generally ratified, it would not, therefore, have established an unequivocal limit to the shelf. In fact, ratifications were slow in coming. The twenty-two necessary to bring it into force were not forthcoming until 1964, and by 1972 only a further twenty-seven states had followed suit (Buzan 1976, p.51). None of the oil-rich states bordering on the Persian Gulf was among them. The importance of the 1958 Convention, insofar as it did carry weight in determining the limits of jurisdiction to be "finally" established by UNCLOS III, was that it clearly supported the contention that any part of a state's shelf, or "natural prolongation", "exploitable" or not, which was shallower than 200 metres, fell under its jurisdiction, regardless of its distance from its shore. It thus correspondingly weakened any subsequent move to base rights to the shelf on distance alone.

Nevertheless between 1958 and 1974, and particularly before 1967, the convention contributed to a process by which coastal states increasingly came to assert rights, not just to the edge of the geological shelf, but to the outer limit of the margin, thus more than doubling (13) the area falling under their jurisdiction. In 1958 "exploitability" was being discussed as if what was at issue was no more than minor variants on a depth limit of 200 metres. By 1967 Canada, in the person of Alan Gotlieb, was responding to Dr. Pardo's initiative by asserting, in the First Committee of the General Assembly, that:

"the present legal position regarding the sovereign rights of the coastal state over the resources of submarine areas extending at least to the abyssal depths is not in dispute" (Buzan and Middlemiss 1977, p.16).(14)

The chief factor accounting for such a rapid change in attitude appears to be the rate of technological advance. What was exploitable, or at any rate explorable, in the nineteen-sixties went far beyond what was thought of as exploitable in 1958. Canada herself began issuing exploration permits for its shelf in 1960, and for its slope in 1963, and, by 1970, even for its rise. Even though these permits on the slope and rise were cautiously drafted as "subject to the lands involved being Canada lands" (Buzan and Middlemiss 1977, p.18), this created a de facto commitment to a liberal interpretation of coastal State claims.

Canada's interpretation did not go undisputed, then or since. For instance, in reply to the Secretary-General's invitation to comment on the Pardo proposal, Malagasy, in January 1968, expressed the view that the shelf should not extend beyond a depth of 200 metres. Coming from a largish state, if not a rich or powerful one, this opinion was not insignificant; but in the main, there were no weighty interests opposed to such dramatic extensions. Even the major maritime powers, concerned about navigation, or distant-water fishing, were reassured by the provision of Article 3 of the Convention that the legal status of the waters and airspace above the shelf were unaffected. That left, as potential opponents, only those advanced countries and their companies that might conceivably benefit from making sea-bed exploitation a "high seas freedom". In fact, however, from the Truman Proclamation onwards, American oil companies took the opposite point of view; they preferred a consolidation of America's rights over its extensive shelf and margin to an assertion of the freedom to exploit the shelves and margins off other countries without their Governments' consent.

This trend was reinforced by the judgement of the International Court of Justice in the North Sea Continental Shelf Cases in 1969, which held that, even in the absence of any treaty obligation:

"the rights of the coastal state in respect to the area of continental shelf that constitutes a natural prolongation of its land territory into and under the sea exist ipso facto and ab initio by virtue of its sovereignty over the land, and as an extension of it in an exercise of sovereign rights for the purpose of exploring the sea-bed and exploiting its natural resources. In short, there is here an inherent right" (ICJ Reports 1969, p.22, cited in Johnson 1969, pp. 531-532).

The phrases "natural prolongation" and "inherent right" were seized on as proof of the legitimacy of coastal state claims to the entire continental margin, though it is by no means clear that the judgement supports that claim. (15)

From the late sixties those that proposed that limits should be based on sea-bed characteristics thus largely spoke in terms of the outer edge of the margin. There were, however, some notable exceptions; the Soviet Union continued to favour a depth limit of 500 metres; the seven land-locked and geographically-disadvantaged authors of A/AC 138/55 in 1971 suggested 200 metres as the depth element (in combination with a 40-mile distance limit); and there were several suggestions for a "blurred" limit, within which the proposed international sea-bed authority would gain a portion of the revenues derived from exploitation. The most striking of these was the Draft United Nations Convention on the International Sea-bed Area proposed by the United States on August 3 1970. It proposed to create an International Trusteeship Area, between the 200-metre isobath and a line to be determined by gradient, "beyond the base of the continental slope", within which, while jurisdiction would be assigned to the coastal state, it would retain only a proportion (between one-third and one-half) of the fees and payments made to it, the rest going to the Authority.

This was a remarkably generous proposal, but it was summarily dismissed by the developing states, possibly because of the adverse connotation of the word "trusteeship". (16)

After this cool reception it was quietly dropped, and at Caracas, when the United States again declared its support for revenue-sharing, though no figures were mentioned publicly, it was privately made clear that such sharing would be on nothing like the lavish scale of the 1970 draft convention. Even so, the US position was still more generous than that of any other proponent of revenue-sharing that might expect to be among the contributors. Some wide-margin states, such as India, suggested it might apply to the portion beyond the economic zone. More typical was the position of Britain, which had once endorsed the Trusteeship Area plan, and now echoed the Canadian view of 1967.

Meanwhile, as the concept of the continental shelf, as a basis for coastal state jurisdiction, expanded to cover the entire continental margin, the distance component of limits to such jurisdiction underwent a transformation even more dramatic; and it is to this that we now turn.

Politically, there is one overriding difference, in establishing such limits, between criteria based on distance and those based on sea-bed characteristics. The latter, often introduced by maritime states, are generally intended to apply only to the sea-bed itself; indeed, in some cases, even more restrictively, only to the resources of the sea-bed (including living resources attached to it). As such, they do not challenge most other maritime activity. Claims based on distance criteria, on the other hand, commonly include rights to sea-bed resources among a number of jurisdictions claimed, which may embrace navigation and pollution (as with the traditional concept of the territorial sea) and will in almost all cases also apply to fishing (as with the concept of the exclusive economic zone). For that reason all changes in distance limits are likely to be controversial; and in some cases the limit of coastal jurisdiction over the sea floor may be an incidental element in a claim advanced (and contested) primarily for other reasons.

The origin of such vastly-expanded distance claims, like that of those based on depth and geology, can be traced to the policies of the United States. Franklin Roosevelt clearly seems to have entertained the possibility of such ambitious extensions, even envisaging asserting title of oil reserves up to half-way across the oceans (Hollick 1976); and in 1939 he persuaded his Latin American neighbours to join him in proclaiming, through the Panama Declaration, a Neutrality Zone of between 300 and 1,200 miles in respect of what became the Second World War. (Hjertonsson 1973, p.19) Nevertheless, the first claims over sea-bed resources to a distance limit of more than twelve miles came from Latin American states, in response to the Truman Proclamation of 1945, and in the teeth of United States' protests. Chile proclaimed sovereignty over adjacent waters and underlying sea-bed up to 200 miles on 23 June 1947, Peru did the same on the first of August of that year, and Ecuador something similar on March 6 1951 (Hollick 1981, p.84); and these three claims were consolidated in the Santiago Declaration of 18 August 1952. Meanwhile, some Central American states had followed the Chilean lead, notably El Salvador, which in 1950 claimed what it called a 200-mile territorial sea (Buzan 1976, p.9), but was in practice a two-tier zone; only in the inner zone of twelve miles was fishing exclusively reserved to nationals, and jurisdiction exerted over navigation (Hjertonsson 1973, pp.67-8).

At the first two UN Law of the Sea Conferences, at Geneva in 1958 and 1960, those that maintained 200-mile claims of this kind could have had little hope that these would be generally upheld. The failure of the 1960 conference worked in their favour. Among the proposals before it were one from the USSR for a 12 mile territorial sea, and one from the USA and Canada for "six-plus-six" (a six mile territorial sea with an additional fishing zone, six miles wide). Either, had it been adopted, might have established itself as a general norm; in the event, the lack of any decision made the unconventional stance of the Latin Americans look less illegitimate than it might otherwise have done.

There was an apparent lull between 1958 and 1965, (17) but the dispute between the United States and the 200-milers in Latin America resurfaced in the late sixties. A series of Congressional Acts committed the United States to disputing such claims. The Foreign Assistance Act was amended in 1965, the Naval Ship Loan Extension Act passed in 1967, and the Foreign Military Sales Act amended in 1968, making, in each case, the "assistance", "loans" and "sales" conditional on US fishing vessels not being interfered with on what the USA regarded as "high seas", though with some presidential discretion to waive this condition. On the Latin American side, Ecuador, in November 1966, converted its Santiago Declaration stance into a claim to a 200-mile territorial sea, and Argentina, apparently in response to Soviet fishing close to its shores, declared that its sovereignty extended to 200 miles. Attempts by the United States to negotiate, both in combination with, and after suspension of, the sanctions established by the aforementioned legislation, came to nothing; and Argentina's example was followed in 1970 by her east coast neighbours, Uruguay, and (in a more extreme form) Brazil (Hjertonsson 1973, p.36).

Dr. Pardo's historic speech of November 1967 showed that he was aware of the serious possibility that the spirit of his proposal might be denied by the acceleration of such claims. He saw the danger, and where the trend was leading, and appealed to governments to halt the process, and "establish some form of international jurisdiction and control over the sea-beds and ocean floor, underlying the seas beyond the limits of present national jurisdiction, before events take an irreversible course" (Pardo 1975, p.2).

Those who then claimed 200-mile zones of jurisdiction (of whatever form) might have responded to this appeal in one of two ways. The main raison d'être of their claim, which it can be assumed they wanted to maintain at all costs, was control over fishing. One way of securing this might have been to separate claims to sea-bed resources from other claims, and to agree to freeze the former at some currently acceptable limit, while emphasising that this in no way impaired their 200-mile claims in respect to fishing or other maritime activities. Such a freeze might have been made conditional on their being satisfied that all countries, and particularly the developing, would benefit from the exploitation of the international area. In this way, they would have avoided diminishing the geographical extent of the "common heritage of mankind" and made sure of not antagonising those developing countries for whom this concept was important, whose support they might well have valued, both on other "law of the sea" questions, and in respect of their economic demands, via the Group of 77, at UNCTAD. For some east coast South American states, this could have been at a cost; if, in consequence, the "shelf" had been more restrictively defined, it would have meant the loss of potential oilfields. For the others, the only cost would have been the loss of face that they might feel they had suffered in being prepared to relinquish what they had once claimed.

The alternative, if they were not to be isolated, was to ensure that the question of delimiting claims was postponed for long enough to enable them to persuade the world community of the appropriateness of a 200-mile limit, even though that reduced what was to be in some sense the heritage of all, and to be run for the benefit of mankind as a whole. This was the course they chose, and between 1967 and 1974, pursued with total success and with decisive effect on the fate of Pardo's initiative. It is possible that their choice was influenced by the fact that that initiative apparently coincided with an attempt in 1967 by the USSR, to which the USA had given encouragement, to call a conference for the much narrower purposes of defining the territorial sea, the limits of coastal state fisheries, and the right of passage through straits (Hjertonsson 1973, p.42).(18) This directly attacked the basic reasons for their 200 mile claim. If successful, it might have done what UNCLOS II had failed to do, and established, in a widely ratified convention, a norm at variance with their position. They might, in any case, have resisted anything that might lead to an early discussion of limits to the shelf, but this move fanned the flames of such resistance.

Their success came with extraordinary rapidity. From being a small minority in 1967, those who advocated a 200 mile limit to coastal state jurisdiction over the sea-bed (whether or not they also envisaged this extending to the shelf or margin beyond 200 miles) had, by Caracas, in 1974, come to constitute a majority so overwhelming that it could be said to amount to a consensus. They had achieved the delay they needed, and they had used it to devastating effect.

They secured delay in three ways. First, when Malta proposed to call a conference specifically on the limits and regime of the sea-bed, (19) they succeeded, by 1970, in widening its agenda into that of the comprehensive law of the sea conference that UNCLOS III was to become. (20)

Since such a conference, by the very magnitude of its agenda, would indeed ensure delay, and thus work against implementing Dr. Pardo's vision of the "common heritage", it can reasonably be asked, why did the majority of developing countries, most of which were sympathetic to that vision, so emphatically support, in 1969 and 1970, the calling of such a wide-ranging conference. One answer, borne out by reiterated developing country statements at Caracas, was that the existing law of the sea, such as it was, had been established before most developing countries had been born, that many of them had not even participated in the relatively limited attempts to reform it in 1958 and 1960, and that it thus necessarily seemed to them to favour the old and the rich. A comprehensive review of it therefore appealed to almost all developing countries as a progressive step, proposals to strengthen the hand of coastal states, in general, against maritime states taking precedence over the advantages, from the point of view of the common heritage of mankind, of insisting that coastal states "freeze" their claims.

A further nail in the coffin of any prompt determination of the limits of coastal state jurisdiction came in 1970, when the General Assembly adopted, without opposition, the resolution that became the basis of all subsequent UN negotiations about the Authority, Resolution 2749. Its main thrust is dealt with below, in chapter 7; here we are concerned only with what it had to say about limits. The preamble to it declared that there is an area of the sea-bed beyond the limits of jurisdiction, whose "precise limits ... are yet to be determined", and Article 1, no doubt deliberately, omitted the word "present" from the original title of Dr. Pardo's item (... the sea-bed and ocean floor ... beyond the limits of present national jurisdiction). Not only were the limits yet to be determined; they were not to be based on the limits then claimed by states. (21)

The Latin American "two hundred milers" (22) complemented these devices effectively with action outside United Nations auspices. Even before the passage of the Declaration of Principles in 1970 they had coordinated their own position in the Montevideo Declaration of May 1970. In this Declaration, the only mention of 200-miles was in the preamble, where they noted that, "by reason of conditions peculiar to them", they themselves "extended their sovereignty or exclusive rights of jurisdiction" to that distance; the operative part merely, in effect, asserted the rights of coastal states to decide upon their own limits, including those applying to their jurisdiction over "the natural resources of the sea-bed and of the subsoil of the ocean floor". (23) Three months later, twenty Central and South American states, including these nine, three from the Commonwealth West Indies and the two land-locked, met at Lima, (24) and fourteen (25) of them endorsed a further declaration, accompanied by three resolutions. The Lima Declaration was as vague and permissive as the Montevideo Declaration, but did not specifically mention coastal state jurisdiction over the ocean floor. The resolutions rallied the signatories to the cause of delay, recommending governments to oppose a conference with a limited agenda, and to regard as premature any attempt to establish the limits of jurisdiction over the ocean floor until, in effect, the Sea-bed Committee had negotiated a regime for the international area.

Once the issue had been broadened to that of coastal state jurisdiction generally, or at any rate that over marine resources of all kinds and not just those of the sea-bed, other developing coastal countries became potential allies of the Latin American 200-milers. They too, with their less-advanced economies, were more likely to find the waters off their coasts used by others. There were exceptions, like Liberia and the geographically-disadvantaged Singapore, which were major flag states; but most shared the "coastal state" orientation, and several, by 1967, had zones extending beyond twelve miles (among them South Korea, Sri Lanka, and Pakistan in Asia, and Ghana and Guinea in Africa (Hjertonsson 1973, pp.142-43)); but these were not necessarily intended to apply to sea-bed resources. Pardo's conception of "the common heritage of mankind" was attractive enough to make several of them hesitate before championing wide extensions of coastal jurisdiction in this field. As late as March 1971 India, Sri Lanka and Pakistan all declared that their attitude to "limits" would be determined by the kind of regime for the "common heritage" that emerged from UNCLOS III. As the Sri Lankan delegate put it:

"The extent of the national jurisdiction which Ceylon might claim or recognise would depend on the existence of a viable international authority with comprehensive powers and acceptable decision-making processes. If such an authority were established and offered the prospect of real benefits to the international community and particularly to the developing countries, his Government would be willing to consider relatively narrow limits of national jurisdiction and would hope to be able to persuade the overwhelming majority of countries to that view, whatever their claims. On the other hand, if agreement could only be reached on machinery of limited scope – a registry of claims, for example, which left the international area a prey to unrestrained exploitation or selfish ends – Ceylon might also find itself compelled to espouse selfish ends, to acquire the means of achieving them by private contract and to lay claim to areas of national jurisdiction commensurate with that aspiration". (26)

The third (Lusaka) "summit" of heads of state and government of 53 "Non-Aligned" countries, none of which had endorsed the Lima or Montevideo Declarations, had in September 1970 issued a statement which was also some considerable way from unqualified support for the Latin American position. It welcomed proposals for holding a comprehensive conference, but called for limits to jurisdiction to be arrived at "in the light of the international regime to be established for the area" and suggested, on a note rather at variance with the unilateralist climate of Lima and Montevideo, that the establishment of such limits, which were to be "clear, precise and internationally-accepted" would be a particularly important task of the conference.

It was thus by no means clear, at the end of 1970, how far the Third World as a whole would follow the Latin American 200-milers, and when the Asian-African Legal Consultative Committee considered the law of the sea as a "priority item" at its 1971 session it showed little unity on this question of limits. (27) By June 1971, however, the Council of Ministers of the Organisation of African Unity (henceforth the "OAU") was beginning to put more emphasis on coastal states' rights than on the value of the common heritage of mankind, though without making specific the limit up to which the former should extend. In November of that year a ministerial meeting of the Group of 77 on marine resources at Lima – the only such meeting to be held at that level on that topic – championed the rights of coastal states to "protect and exploit the resources of the sea adjacent to their coasts and of the soil and subsoil thereof, within their limits of national jurisdiction, the establishment of which must take due account of the development and welfare needs of their peoples" (Buzan 1976, p.185). Against this, in August of the same year, three Asian countries (Afghanistan, Nepal and Singapore) had joined with one European socialist state and three western European states in sponsoring the already mentioned proposal for a distance limit of only 40 miles (the average width of the shelf throughout the world) in combination with a depth limit of 200 metres (28); but it was noteworthy that the African and Latin American land-locked were not among the sponsors. By this time, then, there were signs that the Group of 77 alignment, in which the Latin Americans are allies of Africa and Asia, would prevail, on this issue, over the "non-alignment" alignment, in which Latin American positions had shown themselves to diverge quite sharply from those of most African and Asian states. (29) This was facilitated by the replacement of Taiwan by Peking as the representative of China in the United Nations in 1971, and the immediate cooption of the "new" China onto the Sea-bed Committee, since China supported, in essence, the Lima Declaration's position of leaving it to the coastal state to determine its own limits of resource (and other) jurisdiction; after an apparently inconclusive further session of the Asian-African Legal Consultative Committee in January 1972, which nevertheless revealed extensive support for the idea of an economic zone, combining claims to living and non-living resources, the scales were decisively turned by three events in the middle of 1972.

The first of these was the Santo Domingo meeting of June 1972, at which the principle that coastal states were entitled to a "patrimonial" sea, extending to (at least) 200 miles, was proclaimed by ten Caribbean countries (Colombia, Costa Rica, The Dominican Republic, Guatemala, Haiti, Honduras, Mexico, Nicaragua, Trinidad and Venezuela). Five other states (including two "territorialists", El Salvador and Panama) attended but did not sign.

Almost simultaneously, the Yaoundé seminar, attended by representatives of seventeen African states, and observers from several developed ones, resulted in widespread support for the establishment of economic zones extending a uniform (but unspecified) distance beyond a territorial sea of twelve miles. In form, these conclusions were recommendations to the OAU, which was to meet in the following year; in essence, they represented the adoption of a new policy in Africa.

The third of these decisive events, which followed closely from the Yaoundé seminar, was the introduction of new proposals to the Sea-bed Committee by Kenya. (30) Kenya, one of the leaders at Yaoundé, made clear that she envisaged the creation of "Exclusive Economic Zones" extending, where geography permitted, up to 200 miles. The concept won wide backing, not only among African States, but also now from several others that had previously been uncommitted, such as India, Pakistan and Sri Lanka.

The OAU duly endorsed the principle of a 200-mile Economic Zone in 1973. The land-locked states within the continent did not oppose it, because it was accompanied by a provision permitting them to share in the exploitation of the living resources of the zone. In this respect the African concept of an economic zone differed from the Caribbean concept of a patrimonial sea; the latter, originating in a region devoid of land-locked states, conceded nothing to them.

Thus the tide of support, especially among developing countries, for allowing coastal states' claims to 200 miles, was strong, and was to prove irresistible.

Yet in this same year, those who favoured narrow limits - the "land-locked and the geographically-disadvantaged" - were to win one remarkable, though in the end nugatory, victory. They proposed in the General Assembly that the Secretary-General produce a study examining "the extent and economic significance, in terms of resources, of the international area" that would result from each of the various proposals on limits so far submitted: ie. 200 metres, 3,000 metres, 40 miles and 200 miles. This ostensibly unexceptional proposal was eventually carried, by the convincing margin of 52 to 19 with 48 abstentions, in the First Committee, and, even more emphatically, by 69 to 15 with 41 abstentions, in plenary; but the minorities were substantial enough to indicate significant opposition. More dramatically, an amendment proposed by Canada, France and Malta, which would have robbed the proposal of much of its point, came within an inch of success, achieving a tied vote (46 to 46 with 27 abstentions) in the First Committee. Other amendments, by Peru and Kenya, were also narrowly defeated, by 43 to 39 with 37 abstentions, and by 38 to 28 with 48 abstentions. Thus the proposal survived, and led to the study referred to earlier; but it is extraordinary to find Malta among the leaders of those who would have preferred this information not to have been made available to delegates.

Malta had already, in March 1971, expressed a qualified and reasoned support for the 200-mile principle. (Pardo 1975, pp.210-213) Several developed states, too, were beginning at this time to think of distance limits to coastal state jurisdiction that went far beyond the traditional territorial sea. By 1972 both France and New Zealand endorsed the principle of a 200-mile zone, though in the latter case, for fishing (Hjertonsson 1973, p.114); and Canada and Australia also favoured criteria of depth as well as distance, though Canada, like Malta, combined it with a concept of "custodianship" whereby coastal states would pay part of the revenues derived from their zones as a kind of "voluntary tax" to the sea-bed authority. In the next year Norway, too, endorsed the idea of a 200-mile limit, this time in conjunction with a depth limit of 600 metres (Hjertonsson 1973, p.102). Buzan (1976, p.203) mentions four of these developed states (Australia, Canada, New Zealand and Norway), together with Iceland, which had already established its credentials as a coastal state pioneer in the first of its "cod wars" with the UK, as beginning in 1972 to join with like-minded African, Asian and Latin American states to form a "moderate" coastal state group. The "moderation" displayed by this group consisted in the fact that it favoured 200-mile economic zones rather than territorial seas of the same extent, a distinction that had no relevance for the size of the international area or the revenues of a sea-bed authority.

UNCLOS III

At Caracas, as far as sea-bed jurisdiction was concerned, the main developed states exhibited a striking conversion to the coastal state view. The USA, the USSR and the UK all declared themselves ready to accept the 200-mile economic zone on certain conditions, and the conditions had nothing to do with sea-bed resources. Japan accepted a 200-mile limit expressly for sea-bed jurisdiction. Even the Federal Republic of Germany, which, with Italy, argued in general terms against extensions of coastal state jurisdiction, as unduly reducing the international area, declared itself ready to discuss the principle of a 200-mile zone. With such weighty backing, this principle could no longer be seriously contested. Caracas produced no formal instruments of agreement, but it established quite plainly that in any convention to emerge from the conference, coastal states would enjoy exclusive jurisdiction over all mineral resources, indeed all resources of the sea-bed and subsoil, to a distance of 200 miles from the baselines of the territorial sea.

From Caracas onward, then, the main argument over limits was between those who wanted distance - 200 miles - to be the only criterion - a position adopted by the Organisation of African Unity and most of its members - and those who claimed that coastal state jurisdiction should extend further where possible, to a limit defined by depth or other sea-floor features, most commonly the "outer edge of the continental margin". The first signs of a compromise began to emerge at the next session.(31) It was embodied in the Single Negotiating Text (SNT), Part II, produced by Galindo Pohl of El Salvador, chairman of the Second Committee for that session. In essence, it meant that claims to extend to the edge of the margin were upheld, subject to the condition that the sea-bed authority should receive a percentage, not at this stage specified, of the value of mineral resources extracted from any part of the margin lying outside the 200 mile zone.

This compromise arose directly out of the activities of the unofficial group of delegates that met under the chairmanship of Jens Evensen, of Norway, and closely followed the fourth and final revision of the compromise formula that he put before that group on May 6 1975. The endorsements of Evensen's suggestions in the SNT meant that certain alternative solutions, if not entirely ruled out, were thereafter relegated to the status of rank outsiders. These included., on the one hand, coastal state claims to untrammelled enjoyment of the resources of the margin; and on the other, the African bid to restrict claims to the economic zone; the Soviet proposal, maintained thus far, to a strict depth limit, beyond the zone, of 500 metres; and the earlier idea of the United States, Malta and Canada, reiterated, at Caracas itself, only by the first of these, of requiring coastal states to pay something to the Authority in respect of exploitation within the zone. (32) Not all of these proposals were thereby dropped. As late as 1979, the Arab Group, now heirs to the African position, were still proposing a cut-off at 200 miles, and Nepal was resurrecting, under the label of a "Common Heritage Fund" the idea of revenue-sharing from within the zone. But in the main, after 1975, the arguments were over how the margin, when it extended beyond 200 miles, was to be defined, how the payments the coastal state was to make in respect of it were to be calculated, what states, if any, would be exempted, and by what criteria the monies so accruing were to be distributed.

The RSNT of 1976 and the ICNT of 1977 gave increasing precision to the latter obligation. The ICNT (in its Article 8a) put it at nil for the first five years of production and 1% in the sixth, adding a further 1% in each of the following four years and thus reaching 5%, where it would remain, in the tenth. Parallel with this was a change in the criteria to be applied by the Authority in distributing these revenues. By the SNT in addition to being "equitable" these criteria were to take into account the interests and needs of developing countries (Article 69.4); in the corresponding provision of the RSNT (Article 70.4) the phase "particularly in the least developed amongst them" was added. By the ICNT this had become "particularly the least developed and the land-locked amongst them (Article 70.4). (33) Andres Aguilar, chairman of the Second Committee from the fourth session onwards, saw this as in important concession to the land-locked group, as he made clear in reporting to the plenary at the seventh session in 1978.

"I still believe that recognition of the rights invoked by States whose continental shelf extends more than 200 miles, together with the system of payments and contributions provided for in Article 82 of the Composite Text, and a solution of the aspirations of the group of land-locked and geographically-disadvantaged states, constitutes an essential element of the general agreement on the matters referred to the Second Committee" (UNCLOS III, Records Vol X 1978, p.85).

Even on the matter of revenue-sharing, the ICNT was not to be the last word. By the time Aguilar made these comments, the seventh session had already identified the whole question of coastal states' rights and obligations in respect of the resources of the shelf beyond 200 miles as one of the "core issues" meriting the establishment of one of the seven negotiating groups to be created within the province of Committees I and II. Aguilar had been made Chairman of the Negotiating Group 6, specifically charged with resolving this question. At the next session, the eighth, he was to produce "compromise suggestions" (34) perceptibly modifying the substance of the ICNT, which were adopted as revisions "offering a substantially improved prospect of a consensus".(35) These made one change in respect of revenue-sharing: the proportion instead of staying at 5% from the tenth year onwards, was to rise to 7% in the twelfth year and stay at that.

There remained, on this topic, one point of dispute, which persisted even after the second revision; the question of whether any category of developing states should be exempted from having to make such contributions. The SNT (Article 69.3) and the RSNT (36)(Article 70.3) had left it for the Authority to determine. The ICNT (Article 82.3) specifically exempted, from such obligations, any developing countries that were net importers of the minerals in question; but it in no way implied that they were not to be among the recipients. Some developed countries, including the USA, sought to replace this asymmetrical exemption with an option, limited to a fixed number of years, to contract out of the whole scheme, but this proposal was firmly resisted and no change was made in this provision.

The other part of Negotiating Group 6's task was the definition of the margin. Aguilar had made clear, in his introduction to the RSNT, that he recognised a need for it to be more precisely determined. Before that Evensen, in his "compromise formula" of 6th May 1975, had proposed to give coastal states a choice between two modes of fixing the outer limit; one was to be sixty miles seaward of the foot of the slope, the other the outer edge of the continental margin, comprising "the submerged prolongation of the land mass", including "all rocks appertaining to the said land mass, as well as the overlying sediments of the shelf, slope and rise" but not "the rocks and sediments appertaining to the deep ocean floor". This in itself did not restrict

118

coastal states very much, since the latter alternative admitted of a quite elastic interpretation: but it was accompanied, in the Evensen formula, with a provision for the creation of a Continental Shelf Boundary Commission, to which the Authority, as well as any "interested" state, could appeal, and whose decision would have been "final and binding". The idea of such a commission seems to have found little favour at the time, but it reappeared as part of a complicated "informal suggestion" by Ireland in 1978,(37) and a weakened version of it was resurrected by Aguilar in his compromise suggestions which were incorporated in the first revision of the ICNT in 1979. The body was now to be called the Commission on the Limits of the Continental Shelf, and an annex was promised setting out its structure and mode of operation. Coastal states claiming a shelf beyond 200 miles were required to submit information to it, and take its recommendations into account; but could then establish their own "final and binding" limits.(38) The last word thus apparently lay with the coastal states, but in the first part of the ninth session agreement was reported on a strengthening of the Commission's role, by changing "taking into account" to "on the basis of". Curiously, this change did not appear in the second revision of the ICNT issued on April 11 1980; though it did in the third revision later that year and in the two subsequent texts. In any case, the text of what was to become Annex II made it plain that it was not intended that the coastal state could simply disregard the Commission's recommendations, and that, if it disagreed with them it was required "within a reasonable time", to make a further submission to that body.(39) This came some way towards embodying a proposal made by Singapore at the previous session, by which, in effect, the coastal state would have been able to deviate from the Commission's original recommendations only with the latter's consent.

The first option offered to coastal states under the 1975 Evensen formula was more precise, since it simply meant measuring a fixed distance from the foot of the slope, which could be defined in terms of the maximum change in the gradient at its base. This too was taken up by an Irish proposal, at an informal meeting of the Second Committee at the sixth session in 1977, which at the same time gave a similar precision to the option based on sediments, by also relating it to distance from the foot of the slope. By this proposal, the coastal state could choose to count as part of its shelf any point of the sea floor where the thickness of sedimentary rocks was at least 1% of the distance from the foot of the slope. Sedimentary rocks, if thick enough, were thus in themselves treated as manifestations of the continental margin; no allowance was made for there being also "sediments appertaining to the deep ocean floor", as Evensen had put it; and by choosing this formula, some coastal states would be able to claim shelves extending, at points, several hundred miles beyond their economic zones. Nevertheless, the Irish formulae (as they came to be called), though they might be generous to coastal states, did at least have the merit of reducing the vagueness of previous definitions, which seemed to have given such states carte blanche to say where their margins ended.(40)

These formulae were not incorporated into the ICNT but went before Negotiating Group 6 when it began to consider this "hard core" issue at the seventh session in 1978. Also before it were two other proposals; one, already mentioned, by the Arab Group, for a straight 200-mile limit to the shelf; and a new Soviet proposal, fixing a maximum outer limit of 300 miles (ie. 100 miles outside the limits of the economic zone), as well as a minimum outer limit corresponding to that of the economic zone itself. Where the edge of the margin "as determined by geology and geomorphology" fell between these limits, it was to be the boundary. Negotiating Group 6 also had before it, for the first time, a map showing how the Irish (and certain other) formulae would be applied.(41)

This map was to be of some political significance. The idea of producing it had been brought up by Colombia during the sixth session in 1977,(42) and had generated a debate reminiscent of that in the General Assembly in 1972 over the study on the economic implications of the various limits then being proposed. The land-locked and geographically-disadvantaged states had warmly welcomed this idea, while the "margineers" had responded grudgingly, emphasising its cost and the delay it would impose on the conference's deliberations.

After some delay, the Second Committee met again, and approved the proposal, with one important amendment. The map was originally to have shown five lines: the 200-mile limit;(43) the 500-metre isobath, where it extended seaward of this line; the foot of the slope; and the two variants of the Irish formula. To these were to be added, in the amended version, a line which was to show the outer limit of the continental margin. The Secretariat, and the cartographers they commissioned for the job, were thus put in the absurd position of being required to define independently on the map a limit whose inherent indeterminacy the two Irish formulae had been devised to dispel. In producing this line on the map, they did not indicate how they arrived at it. In fact they based it on an article, published in 1959, of which the Law of the Sea Library at the United Nations did not even have a copy.(44) Later, the International Oceanographic Commission, asked to explore the financial and other implications of a larger scale study, and to add to the original map a line illustrating the new Soviet proposal, was sharply critical of the map for "making definitive interpretations of the formula without detailed supporting arguments or justifications" and "sacrificing precision for speed", and thus containing "a number of errors, omissions and wrong evaluations". In particular, it was so sceptical about the possibility of drawing a line depicting the outer edge of the continental margin, because of "the lack of an accepted definition of this term which would be universally applied", that it found it impossible to add a line illustrating the effect of the new Soviet proposal.(45)

In 1978, however, delegates did not have these criticisms of the map before them. They had only the map, with its line purporting to show the outer edge of the margin, which had the effect of suggesting that neither variant of the Irish formula gave coastal states, in aggregate, anything like the whole of their margins, and thus of making that formula look far less extreme than it had done before. In particular, the map made it appear that, by these formulae, Sri Lanka would suffer a loss of marine territory out of proportion to that suffered by other state.

The immediate response of most "margin" states was to insist that the Irish formula was as far as they would go in limiting their claims, and that unless this was agreed to, other concessions they had made to the land-locked and geographically disadvantaged, notably on a compulsory "conciliation procedure" in support of sharing the unharvested surplus of living resources in their economic zones with such states in the same region, might be withdrawn. This threat did not, however, produce agreement at either the Geneva or the New York portions of the seventh session; on the other hand it did not preclude the development of what Aguilar described as a "positive atmosphere".

Sri Lanka, on the other hand, now apparently uniquely penalised, embarked on a campaign to modify the Irish formulae in its favour. Its first proposal would have permitted states to extend their shelf beyond the line given by these formulae so long as the thickness of sediment was no less than at some point on the lines yielded by the formula, where the foot of the slope was, on average, less than a certain distance from the baselines. This was more generous to coastal states in general than any other specific formula hitherto considered. At the 1979 New York session, a much amended version was submitted, couched so as to apply to Sri Lanka alone. In the following spring, rather than revising the text, it was generally agreed to meet Sri Lanka's plea by incorporating an appropriate declaration by the president of the conference, in its Final Act, after the signing of the convention, though the exact content of the declaration was yet to be agreed.

There was a double irony in this episode. The first was that the country that had, from the outset, provided some of the Sea-bed Committee's most far-sighted leadership, and which, as we have seen, had once urged its fellow developing coastal states, and been prepared itself, to suspend claims to extend jurisdiction, until the regime and institutions for the international area had been determined, had now emerged as the only advocate of making the Irish formulae

121

even more generous to coastal states than they were already. The second, and lesser, irony was that, in form at any rate, the declaration uniquely favourable to Sri Lanka would, but for his untimely death in December 1980, have been made by the former leader of its delegation, whose omission from that delegation, in 1978, had caused the conference one of its major deadlocks.

Most criticisms of the Irish formula, however, came from those who saw it as too generous to coastal states. Mention has already been made of the long-standing Arab proposal that the boundary should coincide with that of the Economic Zone, that is, simply 200 miles from the baselines, and the Soviet proposal, new to the seventh session, for an absolute limit of 300 miles from baselines, which was backed, at the session itself, by a display of maps and diagrams demonstrating the uncertainties and ambiguities of the concept of the margin, if not further defined.(46) This pressure produced some results in the compromise suggestions of Aguilar which eventually appeared in the first revision of the ICNT.

The crux of the change was the stipulation that, whichever variant of the Irish formula was used, what could be claimed must be *either* not further than 350 miles from its baselines *or* not more than 100 miles beyond the 2500-metre isobath.

Even this formula was seen by many as having a serious loophole, in that enormous extensions of marine territory would accrue to states situated on oceanic ridges, at no point more than 2,500 metres below sea-level. Among the attempts to close this loophole at New York was another bid to establish a maximum distance limit, this time by Singapore, at 350 miles; and a further Soviet proposal, in which, more accommodatingly towards coastal states, this provision was to apply only to submerged oceanic ridges. Many coastal (and other) states were prepared to accept this, proved that "oceanic ridges" were narrowly defined. At the first part of the ninth session, Aguilar produced another "compromise formula", said to be the fruit of "particularly intense consultations and negotiations" which was incorporated into Article 76 (47) from the second revision onwards. This provided that, "on submarine ridges", the outer limits of the continental shelf should not exceed 350 miles from the baselines, but that this limit was not to apply to "submarine elevations that are natural components of the continental margin such as its plateaux, rises, caps, banks and spurs". Although this fell short of being an "agreed text", it enjoyed wide, if slightly qualified, support.(48) The chief opposition to it came from the LLGDS and the Arab group; the latter, having modified its outright opposition to coastal states' claims to the margin outside the economic zone, now sought, unsuccessfully, (as had the USSR earlier) to require that a single overall limit should apply.(49)

As we have seen, the boundary between national and international areas of the sea-bed is not fully determined by establishing an economic zone of a specified width and unequivocally defining the extent to which coastal states can lay claim to the shelf beyond this zone. There remain questions of how to draw the baselines from which the zone extends and how to treat archipelagoes, small islands and rocks. These questions have not been particularly prominent at UNCLOS III, and were not among the "hard core" issues arising out of the ICNT which were referred to the newly-created "negotiating groups".

The main reason for this was probably the fact that, apart from the issue of archipelagoes, these were all familiar questions. They had had to be at least verbally resolved by UNCLOS I in 1958 in drafting the Convention on the Territorial Sea and the Contiguous Zone. UNCLOS III thus had before it a precedent, though a loose and ambiguous one.

By that convention, a standard of "normality" for baselines, the low water line along the coast, was affirmed (Article 5), but straight baselines were permitted "where the coastline is deeply indented and cut into, or if there is a fringe of islands along the coast in its immediate vicinity" (Article 7.1). Although the drawing of such baselines "was not to depart to any appreciable extent from the general direction of the coast" (Article 7.3) and the sea areas within the lines had to be "sufficiently linked to the land domain to be subject to the regime of internal waters" (Article 7.1), they could be drawn from "appropriate points" (not necessarily points on land) and could take into account "economic interests peculiar to the region concerned, the reality and the importance of which are clearly evidenced by a long usage (Article 7.5)." Moreover, while precise provision was made for straight lines not more than twenty-four miles long to be drawn across bays, 'historic bays' (Article 10.4), or cases where straight baselines were applicable by reason of Article 7, were exempt from such constraints (Article 10.6). Terms like "deeply indented", "immediate vicinity", "any appreciable extent", "appropriate points", "sufficiently linked", "economic interests", "long usage" and "historic bays" are clearly liable to be interpreted in a variety of ways, and coastal states have, in practice, shown little inhibition in adopting interpretations to their maximum advantage.(50)

Even less restrictive criteria for drawing baselines surfaced in the Second Committee at Caracas and were noted in Andres Aguilar's "main trends" on the topic produced at that session. One version proposed to dispense entirely with any attempt to prescribe a "normal" method of drawing base-lines; another would have permitted them to be drawn from low-tide elevations, not only if lighthouses or similar installations had been built on them (as the 1958 Convention permitted) but also if states had "historically and consistently applied them". The SNT of 1975 did not accept all of these, but embodied what have been described (Pardo and Borgese 1976, p.21) as "three further major departures (as compared with 1958) from the supposedly normal rule for baseline drawing": the legitimising of "mixed" baselines (combinations of straight baselines and those based on the actual low-water marks);(51) the use of low-tide elevations to draw straight baselines in instances where this practice has "received general international recognition"(Article 6.4) (a slightly more restrictive condition than that mentioned earlier from the Caracas "main trends" document); and a special provision for unstable low-water marks in the vicinity of deltas (Article 6.1). It also embodied an article on reefs, allowing the baselines, in the case of "islands situated on atolls" or "having fringing reefs", to be "the seaward edge of the reef".(52) In the RSNT and the ICNT (including the latter's first two revisions) there is virtually no change of substance in these provisions. Pardo and Borgese suggested setting a maximum length for straight baselines, requiring them to be drawn from land points only, abolishing the provisions relating to deltas, gradually eliminating the concept of 'historic bays' and allowing the international authority the right to challenge the validity of any coastal state claim before an independent body. The only one of these suggestions to be even partly implemented was the last, in the form of the provision for establishing a Commission on the Limits of the Continental Shelf, referred to earlier, which was among the "compromise suggestions" Aguilar introduced in 1979, but that commission was presumably to be confined to considering claims extending beyond the 200-mile zone.(53)

The convention, and the texts which preceded it, thus leave it quite uncertain where baselines should be drawn. This fact helps to explain why in 1978 the Secretariat, in producing a map showing, among other things, a 200-mile line, prudently refrained from using baselines at all. How much this uncertainty will affect the size of the international area depends on how long the longest baselines that result from it are. Straight baselines up to a hundred miles long will not much affect the extent of a 200-mile zone, since circles of 200-miles radius drawn from their endpoints will intersect less than seven miles short of the line drawn from the straight baseline, and include all but at most 200 square miles of the area straight baselines would yield. A straight baseline 200 miles long would be more serious, and could transfer something in the region of 2,500 square miles from international to coastal state jurisdiction.

These mathematical implications are also, of course, relevant to the rules for drawing the baselines of archipelagoes. The doctrine that archipelagic states were entitled to use, for this purpose, lines joining the outermost points of their outermost islands, was apparently first proclaimed by Indonesia in 1957 (Luard 1974, p.159). By 1973 Indonesia, with three other leading exponents of "the archipelagic principle", (Fiji, Mauritius and the Philippines) had elaborated this doctrine in two papers to the Sea-bed Committee,(54) and a proposal based on these was submitted to the Second Committee by these same four states.(55) This included the phrase the "outermost... drying reefs of the archipelago" and thus did not, even, require the lines to be drawn from points on "islands" as such; and it contained no suggestion for limiting the length of baselines or the proportion of sea to land in the area they enclosed. In contrast, a British proposal of 1973 (56) would have set a limit of 48 miles to the length of baselines, and have established a maximum ratio of sea to land within them of five to one.

The SNT embodied much of the language of the Caracas proposal of the four archipelagic states, including the apparently self-contradictory definition of such a state as one "constituted wholly by one or more archipelagoes and may include other islands".(57) Baselines could also be drawn to connect the outermost points of drying reefs as well as islands. But, in line with the British proposal of 1973, though with greater leniency towards archipelagic states in the figures used, limits were set on the length of baselines and the ratio of sea area to land area they might include. The latter was put at nine to one; the former at 80 miles, but allowing an unspecified percentage to extend to 125 miles. In the RSNT this became 1%. The ICNT of 1977, however, which in this respect proved to contain the eventual wording of the convention, considerably relaxed the stringency of these constraints. Three per cent of baselines were now to be permitted to be as long as 125 miles, and the limits for the length of the remainder became 100 rather than 80 miles. Nevertheless, even these more permissive provisions would not, as we have seen, transfer much from the international area to the economic zones of such states, compared with baselines drawn from the same points that more closely followed the line of the coast.

The one element of the boundary of the international area which seems to have been effectively resolved, with an ease, not to say casualness, totally disproportionate to the magnitude of what was at stake, was the treatment of islands. Momentous though this issue was for the common heritage, a form of words for it was articulated early in the proceedings, and all subsequent texts have left the original formula unchanged, according all islands, regardless of size, the same rights to economic zones and continental shelves as mainland territories, with one proviso, that " rocks which cannot sustain human habitation or economic life of their own shall have no Exclusive Economic Zone or Continental Shelf."(58) This exception was contested with great tenacity, notably by the UK, mindful of the gains that could accrue to it from an economic zone and continental shelf attached to Rockall, which can hardly fail to fall in to the category debarred; but though, after the text had been formalised, a formal amendment to this effect was proposed at the final session in the spring of 1982, it was, like most other such formal amendments, not pressed to a vote, and not thereafter seen as a sticking-point in negotiations.

The main reason why possessors of islands have, in general, been treated so generously at UNCLOS III seems to have been the effect of precedent. The 1958 Convention gave islands (defined, not in terms of habitation or potential economic life, but merely as "naturally-formed areas of land surrounded by water" which are " above water at high-tide") the same rights to a territorial sea as other land territories. Since, at that time, territorial seas were the main zones within which coastal states enjoyed resource jurisdiction, it was logical, if inequitable, to apply the same principle to the vast tracts that were to become economic zones under the new convention.

Not all advocates of exclusive economic zones, however, accepted this bleak logic. The OAU, for instance, after meetings of its Foreign Ministers at Addis Ababa in 1973 and Mogadishu in 1974, declared that "determination of the nature of maritime space of islands ... should be made according to equitable principles taking account of all relevant factors and special circumstances"(59) among which, it was suggested, were size, population, contiguity, geology and the special interest of island states and archipelagic states.

Even island members of the OAU, such as Madagascar and Mauritius, supported this position in the Second Committee at Caracas.(60) Other states, such as Malta and Romania,(61) put forward other proposals for distinguishing genuine "islands" from "islets" and (in the case of the Romanian proposal) "islands similar to islets".

Nevertheless, all further attempts to restrict the rights of owners of islands (others than those under foreign and colonial domination) came to nothing. There seem to have been two main reason for this. One was the extreme difficulty of devising an objective definition of an inhabited island that did not result in an arbitrary and embarrassing distinction being drawn between those fully endowed with economic zones and continental shelves, and those not. The other was a general sympathy for the predicament of the small, isolated, and not very populous, oceanic state, for whom the assurance of an economic zone, free from the possibility of exploitation by outsiders, seemed little more than proper compensation for a genuine, if paradoxical, form of "geographical disadvantage". This feeling tended to spring from considerations of fishing: it ignored the inequities (and social disruption) springing from giving such small islands such extensive rights over rich sea-bed resources at the expense of the world as a whole: and the incapacity of such tiny states to effectively administer the enormous areas they would acquire. It also ignored the fact that many small isolated islands belonged to continental coastal states and merely added such bounty as they brought with them to states not disadvantaged at all (in many cases) and already rich in sea-bed territory. Nevertheless, these considerations, added to the momentum generated by the 1958 precedent, produced an outcome which permitted further enormous chunks to be carved out of the international area, in addition to what had already been lost to the economic zone and the principle of "natural prolongation".

UNCLOS III thus continued, for the sea-bed as well as the seas themselves, the process of 'enclosure' that had begun with the Truman Proclamations of 1945. The possibility that what was left unenclosed might become "the common heritage of mankind" has not arrested this trend. The convention made it clear that what would be left to the international authority would be a severely attenuated version of what Pardo could have envisaged in 1967, if the limits of national jurisdiction had then been frozen. That does not mean that it was in consequence, economically insignificant. Coastal states had taken much, in one way or another, from the common pool of the world's mineral resources since 1945 and, indeed, even since 1967; but they had not taken by any means everything. Vast tracts remained international. Even if manganese nodules proved to be their only valuable contents (and it now looks as if they will not), the task of creating a system to govern their exploitation was far from devoid of economic, as well as political, moment.

NOTES

1. It has been estimated that, if every part of the sea-bed, as it became exploitable, fell under the jurisdiction of the nearest coastal state, three-quarters of the whole would accrue to eighteen states, and two-thirds, to a mere eleven (Pardo 1978, p.198).

2. What has been disputed, of course, is how wide a territorial sea can be claimed, and, as will be explained below, on what principles its baselines may be drawn.

3. The rationale of "blurred" limits will not be separately considered in general terms, but certain proposals for "blurring" these limits in specific ways will be looked at later.

4. see above, chapter 2.

5. This last argument is somewhat weakened by the success with which the claimants to the North Sea shelf conducted their operations in what, until recently, was high seas, without apparently engendering any major argument with other sea users, and also by the possibility that sub-sea completion systems, which must constitute much less of a

traffic hazard, will replace drilling platforms, at least in deeper waters. See above, chapter 2.

6. The language is taken from Article 4 of the 1958 Convention on the Territorial Sea.

7. Prescott 1975, Table 2, p.80. The three were Argentina (105 miles), Burma (75) and Ecuador (52): the other three making up the original six were Guinea, Haiti and Madagascar.

8. The language is taken from article 4 of the 1958 Convention on the Territorial sea.

9. An even longer list of possible ambiguities, mostly arising out of distance limits, can be found in Pardo and Borgese 1976, pp.21-28. The devastating analysis of this paper makes it even more difficult to understand a passage in an earlier speech of Pardo's, in which he concluded that "there is no alternative to selecting the criterion of distance from the coast to define the outer limits of the jurisdiction of the coastal state in ocean space", giving as one of his reasons "the difficulty of definition" (of the alternative of sea-floor characteristics) and even going on to assert that "a general two hundred mile maximum outer limit of coastal state jurisdiction ... would be sufficiently precise to permit international institutions to start working usefully" (Speech to the Sea-bed Committee, March 23, 1977, in Pardo 1975, p.212). Even in that speech, though, Pardo is well aware of the above problems.

10. Pardo (1975, p.209) goes further and argues that such a doctrine would permit Germany to incorporate Denmark, (or, one might add, Argentina the Falkland Islands). It can hardly be disputed that, as applied to an inhabited territory, it disregards the principle of self-determination, and therefore can lead to clear injustice. Applied to an uninhabited tract of sea-bed, though, however unpersuasive it may be, it could hardly be condemned on those grounds.

11. Yearbook of the International Law Commission 1950, Vol. 1, summary records of the second session, June 5 - July 29 1950, 67th meeting, pp.218 and 212, cited in Sreenivasa Rao 1975, p.50 and notes 19 and 20 on p. 226.

12. Mr. Garcia Amador later denied ever having made such a statement.

13. From 22 million sq.km. to 45 million sq.km.

14. Ironically, in 1958, Canada had supported the exploitability criterion only reluctantly, after failing in a series of moves aimed at specifying the limits of the shelf more precisely (Buzan 1976, p.39).

15. The cases it had to decide related to an expanse of seabed, below the North Sea, nowhere more than 200 metres deep; the judgement repeatedly refers to the continental shelf, defined at one point as "an area physically extending the territory of most Coastal States into a species of platform" (which hardly supports claims to the continental slope, much less the rise); and a good deal of weight is put on the notion of "adjacency", so that at one point it is observed that a point some 80 miles from the coast of Denmark could not possibly be deemed adjacent to it.

16. A significant indication of developing country reaction to this U.S. proposal came from the 1971 meeting of the Asian-African Legal Consultative Committee, where it was criticised for allegedly permitting the "trustee" to gain more than the "beneficiary", and for contradicting the common heritage principle, criticisms which implied support for narrow limits of jurisdiction, at least as against claims based on natural prolongation, or other sea-floor characteristics. See Buzan 1976, p.184 and n.8.

17. Hjertonsson (1973, p.30) is inclined to attribute this lull to a de facto "non-enforcement" policy on the part of Ecuador and Peru in this period.

18. See also chapter 5 above.

19. A/AC 138/11, described in Buzan 1976, p.96.

20. The final decision embodying this came in General Assembly Resolution 2750C, quoted in Buzan 1976, p.112. See also chapter 5 above.

21. There is another irony here. Earlier, Peru had objected to the dropping of the word "present" from this phrase, apparently believing that without it "the limits of jurisdiction" could be taken to mean current norms, at variance with the Latin American claims (Mitchell 1974).

22. By then consisting of Argentina, Brazil, Ecuador, El Salvador, Nicaragua, Panama, Peru and Uruguay.

23. Montevideo Declaration, quoted in Oda 1972, p.348.

24. There were also observers from Canada, South Korea, Costa Rica, India, Ireland, U.A.R., Senegal and Yugoslavia (Hjertonsson 1973, p.71).

25. The British Commonwealth states, the land-locked, and Venezuela, were the exceptions (ibid.).

26. A/AC 138/SR47, p. 2, March 15 1971, quoted in Alexander 1971, p.6. Ironically, Sri Lanka was subsequently to claim for itself a unique relaxation of the already generous criteria which were to find favour at UNCLOS III. See below.

27. This session was apparently the scene of intense diplomatic activity by the Latin American coastal states and thus a crucial stage in their campaign to convert the developing world as a whole to the side of wide extensions of coastal state jurisdiction. See Buzan 1976, p.184.

28. A/AC 138/55 of August 20 1971.

29. For a statistical demonstration of this divergence, on the cold war and on relations with Southern Africa, see Willetts 1978.

30. A/AC 138/SCII/L10.

31. i.e. the Third, at Geneva in 1975.

32. The US had suggested, at Caracas, that such sharing should begin at 12 miles or 200 metres (Ogley 1974, p.616).

33. Emphasis mine.

34. This may have been in response to Nepal's "Common Heritage Fund" campaign, referred to earlier, and to a proposal from Sri Lanka, which would have compensated the "common heritage" somewhat by introducing a much more onerous scheme (NG 6/6 of 10 April 1979) for revenue-sharing from outside the zone, beginning with 4% of the value of production from such areas in the first five years of a mining operation to 8% in the next five, 17% in the next ten and 15% thereafter. Three categories of developing countries (the Least Developed, the Most Seriously Affected, and those that were net importers of the products in question, or did not use them in international trade) would be exempt from such obligations; other developing countries were to pay one quarter of the above percentages.

35. A/Conf. 62/L37 of 26 April 1979.

36. The RSNT, unlike the SNT, would have given the Authority the right to decide if (as well as to what extent) such countries should be exempted.

37. NG 6/1 of May 1 1978, para 5.

38. A/Conf. 62/L37. Article 76. para 5.

39. A/Conf. 62/W.P. 10/Rev. 2, Annex II, Article 8, which remained, though Rev. 3 and L78, and the eventual convention.

40. Exactly how far the vagueness of the concept had been thus reduced depended on the precision with which "the foot of the slope" could be measured. It has been argued (Sreenivasa Rao 1975, pp.8-9) that this is difficult to identify, but it is undoubtedly less so than the line between the rise and the abyssal floor, even in cases where

the standard pattern of transition was followed (see chapter 2 above).

41. A/Conf. 62/C.2/198.

42. UNCLOS III Records, Second Committee, 50th meeting, 23 June 1977.

43. This as we have seen, could easily have involved the cartographer in the delicate and potentially acrimonious business of pronouncing on baselines; but they neatly sidestepped the whole question, by confining themselves, in the drawing of this line, to points 200 miles from land, and explaining that the use of baselines would make the actual extent of exclusive zones somewhat greater than this.

44. i.e. Heezen, Tharp, and Ewing 1959. The cartographers, like the authors of this article, were associates with the Lamont Geological Observatory of Columbia University, New York.

45. A/Conf. 62/C.2/L99, paras 6-8.

46. There is more irony here in that, as we have seen, it was for this very reason that the IOC was unable to add to the Secretariat map a line indicating the effect of the Soviet proposal.

47. In the form of a new para. 6 and an amendment to para. 3.

48. The USA, for instance, supported it on the understanding that "features such as the Chukci Plateau and its component elevations to the north of Alaska" would not count as a ridge (US Delegation Report, 1980, p.31).

49. See UN Press Release, SEA/396 4 April 1980, p.29.

50. Examples of the extravagance with which such states have enlarged their 'international waters' by drawing straight baselines have been given above. Most Latin American states, particularly addicted to this practice, were not parties to the Convention on the Territorial Sea. But of the six states which included in their baselines a leg exceeding a hundred miles in length, two, Haiti and Madagascar, were.

51. SNT Part II, Article 6.2, 7 May 1975, Article 6.2.

52. SNT Part II, Article 5. From the RSNT onwards, this became the 'seaward low-water line of the reef'.

53. Since one of the criteria for delimiting the shelf beyond the Exclusive Economice Zone referred to distance from the baselines from which the territorial sea is measured, it might be open to the Commission to question the

baselines drawn by a coastal state in justifying a claim of this kind; though such a challenge would require some temerity.

54. A/AC 138/SCII/L15 in the spring 1973 session, and A/AC 138/SCII/L48 in the summer session.

55. A/Conf. 62/C2/L49 of August 9 1974.

56. A/AC 138/SCII/L44 in the summer 1973 session.

57. SNT, Part II, Art. 117.2(a). In spite of the seeming contradiction this definition does clearly imply that states with continental territory cannot claim to be archipelagic states merely by virtue of possessing archipelagoes. The next article further restricted the category of such states by also specifying, within the baselines so drawn, a _minimum_ ratio of water to land (at least one to one).

58. SNT Part II, Article 132.3; eventually Article 121.3 of the convention.

59. A/Conf. 62/33 of July 1974, para 5.

60. They were among the sponsors of A/Conf. 62/C.2/L62 of August 14 1979, a series of draft articles on the regime of islands which would have distinguished "islands" from "islets", "rocks" and "low-tide elevations", and further classified all these phenomena into "adjacent and non-adjacent". The gist of the proposal was that states would not be able to claim jurisdiction over marine space by virtue of sovereignty or control over non-adjacent islets, rocks or low-tide elevations, though it could establish safety zones "of reasonable breadth" around them, and that the zone claimable in respect of non-adjacent _islands_ would be determined by "equitable criteria" including size, geographical situation, and "needs and interests" of their population. The distinctions, however, were loosely made ("islands" were "vast naturally formed areas of land" as opposed to "islets" which were "smaller naturally formed areas of land"; "adjacent" was defined as "in the proximity of") and seemed as much concerned with the vexed question of delimitation between opposite and adjacent states as with preserving the common heritage from the encroachment of coastal states, though, had the definitions been given precise content, they would have had this effect.

61. Romania's proposal (A/Conf. 62/C.2/L53 of August 12 1974) defined "islets" by area (i.e. less than one square kilometre), and created a category of "islands similar to islets" which could be somewhat larger but were not permanently inhabited and did not have an economic life of their own. Neither of these kinds of entities were to give rise to claims to marine space, other than territorial waters or 'security zones'. Again, in this proposal, as in the African one, there is much emphasis on the implication

for delimitation between neighbouring states.

7 The System of Exploitation

This chapter covers the main substantive issue with which
UNCLOS III, in attempting to procure agreement on the
establishment of a sea-bed authority, had to deal. It is
therefore a long chapter, though it could easily have been
longer. Three relatively self-contained aspects of this
central issue (production control; the structure of the
authority; and the system of dispute settlement) have been
given chapters of their own. So, too, has the more recent
issue, posed by unilateral legislation, in the USA and
elsewhere, from 1980 onwards, as to how investment
authorised nationally was to be reconciled with the system
of the convention, assuming that the latter was eventually
ratified by all the authorising states. It would be
difficult to disaggregate further what remains. The system
of exploitation as here conceived has to be seen as a whole;
and indeed, to make it intelligible, reference has to be
frequently made, even at the cost of some repetition, to
what was happening with respect to the issues dealt with
more fully in other chapters.

The chapter will do four things: first, it will consider
objections raised against not merely the concept of the
"common heritage of mankind", but the very necessity of any
international regulation of sea-bed mining. Secondly, it
will describe the vast initial gulf between developed and
developing states, with respect to the form this
international regulation should take and the evolution of
the idea of the parallel system by which it seemed in 1980
it was to be bridged. Thirdly, it will examine closely the
terms of this quasi-agreement of 1980 as embodied in the
draft convention of that year. Finally, it will chronicle
the crisis precipitated by the Reagan review of 1981 and the
consequent demands for change emanating from the US
delegation (though in many cases with the support of other
states) with which the eleventh session at New York, in the
spring of 1982, tried unsuccessfully to cope.

THE NEED FOR REGULATION.

> "When the miner goes beneath the sea and beyond the
> territorial jurisdiction of a State, he encounters a
> new set of forces. One is the amorphous but
> formidable jellyfish consisting of the general body

of International Law. The other is a ravenous shark
attracted by the miner's first tentative undersea
movements. It might aptly bear the sobriquet of
"Jaws", but is known more formally as the
International Sea-bed Resource Authority"
(Ely 1975, p.3).

"Inefficiency in ocean mining is least likely to
occur without the property arrangements that appear
to promote efficiency for [other] ocean resources"
(Eckert 1978, p.251).

These two quotations show that there are some who, far
from accepting the validity of the claim that the resources
of the international area of the sea-bed should be treated
as "the common heritage of mankind", do not even see any
need for international regulation. The two analysts,
however, differ in what they prescribe. Ely, in the speech
quoted, was defending the claim of Deepsea Ventures, in
1974, to the exclusive right to mine a "deposit" extending
to 60,000 square kilometres of the Pacific, the right being
derived from discovery. Eckert denies the necessity for any
such exclusive rights, whether based on "discovery" or on
the award of a national or international regulatory body,
and regards them as conducive to inefficiency.

Eckert thinks in economic terms, and emphasises the
criterion of efficiency. He also assumes that sea-bed
mining would be confined to nodules and that high-grade
sites are abundant. After considering the question of
whether, in the absence of exclusive title to a site, an
initial investment on a given site would attract "poaching"
competitors, he concludes that, for a variety of reasons, it
will be more profitable for a mining company to explore new
deposits than to "poach", and that if it did poach, the
initial investor could easily "buy off" the poacher. Much
of the argument hinges on the claim that investment in the
mining machinery necessary to exploit a deposit is specific
to that deposit and requires full exploration of it; and
thus the second investor could compete successfully with the
first in the same deposit only by incurring the same
exploration and development costs (assuming the first
investor is able to keep the results of its explorations,
and its technology, secret).

The implausibilities of this argument are many. As was
argued in chapter 2, it seems unlikely that expensive
equipment would be designed for only one deposit, or that a
"deposit" would be sufficiently homogeneous in itself and
distinct from all other "deposits" to warrant such fine
tuning. It also seems unlikely that the best sites will be
so abundant that they will not attract considerable
competition; and even if "poaching" were to be unattractive
in itself, it might become attractive if it was to evoke the
payment of "inducements to desist" by the original
investor. (1) Moreover, as Eckert concedes, a "poaching"

company can "poach" staff as well as terrain; the original investor could lose the trade secrets on which his advantage, in the site at issue, depends. It is not difficult to envisage a situation where, even if fairly small scale investment was possible, the small investor could quickly be put out of business by large-scale consortia exploiting economies to scale in mining equipment and using the same equipment, profitably if not perhaps with maximum efficiency, on a variety of sites, and denying any reward to the prospecting and exploration contributed by the smaller pioneer.

These, moreover, are only the economic arguments. Add the possibility that some of the other 150 or so states in the world might have political or strategic motives for engaging in mining, or other activities, in a part of the international area in which some investor was exploiting a non-exclusive right to mine, and the hazards are compounded. Thus, even if no consideration is given to the interests of mankind as a whole, law is essential if peaceable, profitable exploitation'of the international sea-bed is to occur; but what kind of law?

The principle that Ely recommends, of simply acknowledging as legally valid claims to exclusive title made by non-state actors, would constitute an odd kind of law. Legal claims need to be scrutinised and validated within some system of rules. If it is open to the claimant to delimit the area claimed, for instance, it would equally open for one claimant to claim the entire sea-bed, or at least, an entire province believed rich in potential. Somebody other than the claimant, then, needs to prescribe how large a site should be. Many would argue that this can be the claimant's own state; hence the rationale for legislation in the USA and elsewhere. But given that the area claimed is undoubtedly international, no other state need recognise a claim granted by a national law unless it commits itself to do so. Different national authorities could thereby allocate coinciding or overlapping sites, unless they provide for some system of mutual recognition, and if different countries issue licences on different terms, mutual recognition cannot be taken for granted. For instance, a less-advanced state may wish to allocate a site to itself, or one of its nationals, even though it cannot exploit it for some time; a coastal state may be prepared to extend its jurisdiction to contest a site claimed under another state's jurisdiction; a state less concerned with environmental hazards may impose less strict rules than another, and thereby attract registration of mining enterprises under "flags of convenience". In the absence of universal agreement as to the law by which legal title may be issued, the investor is exposed to legal uncertainty and perhaps arbitrary political action. At the very least, efficiency requires a registration of claims; and, if nodule sites are not so abundant as to be as free as air, efficiency also requires that, to promote their efficient use, claimants should pay competitive rents, and insofar as there is general agreement as to how mining should proceed, there is a case for establishing a body able to monitor compliance with these agreed standards, so long as administrative costs do not absorb the benefits derived therefrom. Thus on grounds of economic efficiency alone, a sea-bed authority is necessary.

Thus the rhetoric of the "common heritage of mankind" has a firmly practical base. What that phrase does is not to invent the necessity for global action, but to give it such vivid expression as to prompt questions about what other functions a sea-bed authority might have. If the sea-bed, or, at any rate, this tract of it, is regarded as belonging to mankind as a whole and if the Sea-bed Authority can be regarded as representing mankind, then should not the revenues it receives be used to diminish, if only to a tiny extent, the international inequalities of wealth with which this present world is afflicted? Should the exploitation of the sea-bed not be so regulated as to make it a vehicle for transferring technology from the advanced countries to the less advanced? Should not this new institution reserve for itself the right to mine this heritage of which it is appointed the trustee; or, if it is to assign states and companies a role in that heritage's exploitation, should it not have unrestricted discretion to control the rate of exploitation and to say who should exploit it and under what conditions?

It is when we pass to these questions that we move into the area of greatest controversy. It is with these aspects of the system of exploitation that the negotiations of UNCLOS III and the Sea-bed Committee have been largely preoccupied.

On the one side have been those that have tried to confine the functions of a Sea-bed Authority to the demonstrably necessary, or even less, and have not been afraid to imply that if UNCLOS III insists on vesting it with what they regard as excessive powers, they will contrive to provide whatever legal backing is needed for sea-bed mining outside that forum. On the other are those that argue that, because the Sea-bed Authority is the instrument by which the common heritage of mankind is to be realised, it needs to be given all the functions listed in the preceding paragraph, and perhaps more. Additionally, and closely-related to the question of what should be the Authority's functions, lie the questions of how plausibly it can be said to represent the "mankind" whose heritage it will be asked to administer, and how efficiently it can be expected to act on its beneficiary's behalf.

The fundamental dilemma, then, in these negotiations, has been this:

The dangers and potential for damage and injuries arising out of ocean mining are serious enough, and the present state of knowledge so patchy and unreliable, that what is needed is a world-wide body which can not only enforce such regulations as states might now agree to in a convention setting it up, but can also revise the framework within which such mining is conducted in response to experience and the development of new knowledge. On the other hand, the body thus created will itself be novel, and without close precedent in global organisation. To the extent that, like those of the General Assembly, its decisions reflect the numerical preponderance of developing countries, it is seen as hostile, or at least unsympathetic, towards the developed world, and towards foreign investment; and although only a few developing (and a few developed) countries will be directly affected by the competition the sea-bed may constitute for land-based producers, the moral support the developing countries as a whole have given to oil-exporters has suggested to many in the developed states that a body controlled by the former might and will be disposed to obstruct sea-bed mining, if not to strangle it altogether. Thus the need for a world-wide body to regulate (and tax) sea-bed mining in the global interest is balanced by uncertainty as to how such a body will interpret that global interest. Because this is indeed a genuine dilemma, the debate in the Sea-bed Committee's First Subcommittee, and subsequently the conference's First Committee, at its best, has been characterised by a profound and absorbing dialectic.

THE PROCESS OF ESTABLISHING A PARALLEL SYSTEM.

When Dr. Pardo first raised the sea-bed question in the General Assembly in 1967, he envisaged the eventual creation of an agency with "wide powers", able to grant rights to explore and exploit petroleum as well as nodules, and to control ocean pollution. This "long-term objective", as he called it, was to be achieved by 1970. In the event, as we have seen, (2) it took six years of deliberations in the Sea-bed Committee (and the General Assembly) before even a conference could be convened. The major landmarks of these six years were two (among many) resolutions of the General Assembly (the highly controversial Moratorium Resolution of 1969, (3) and the unopposed Declaration of Principles of 1970, (4)) and twelve proposals from specific states or groups of states, including two from Britain, put before the Sea-bed Committee in 1970 and 1971."(5)

The Moratorium Resolution of 1969 barely managed to win a two-thirds majority (the voting being 62 for, 28 against, and 28 abstentions) and was the only occasion in the entire history of the issue where the developed states were outvoted en masse by a developing country majority. (6) The thrust of the resolution was to affirm the conclusions of the preceding section of this chapter: no exploitation of the international area was to take place until a regime had been established. Although those who opposed the resolution were quick to dismiss its impact and emphasise that it was not legally binding, it may well have discouraged early investment in sea-bed mining by casting some doubt upon its legal status in the absence of an internationally-agreed regime.

The Declaration of Principles, carried by 108 to nil with fourteen abstentions, has also been regarded as, eventually, a set-back for the "common heritage":

> "While this accomplishment superficially appeared to
> be a success for Pardo, the underlying reality
> looked much more like failure" (Buzan 1976, p.110).

The most powerful ground for this judgement was the fact, discussed in the previous chapter, that it explicitly avoided defining the limits of coastal state jurisdiction, beyond affirming that there was an area that lay outside it and was international - a valid criticism; but with respect to what happened within this international area, it cannot be denied that it set the framework, subsequently accepted even by those that abstained on it, within which discussion was henceforth to proceed.

Much of the declaration was lofty and vague; but not all. It not only designated the international area as "the common heritage of mankind" (para.1); it called for the creation of an international regime "including appropriate international machinery" by means of "an international treaty of a universal character, generally agreed on", one aspect of which would be provision to ensure that all states, particularly the developing countries "whether land-locked, or coastal" should share in the benefits derived from ocean mining (para.9). Echoing the Moratorium Resolution, it enjoined that all exploration and exploitation of the Area and other related activities would be governed by the regime (para.4); but it also restricted the latter's scope so as to leave legally unaffected the superjacent waters and air space (para.13a).

Some choices had thus been made which would help to shape the agenda of the First Committee of UNCLOS III and the First Sub-Committee that preceded it; though the declaration's wording permitted ample scope for divergent interpretations. The twelve proposals of 1970 and 1971 show just how wide a range of views still persisted, particularly in respect of coverage, redistributive potential, and above all, the power to be invested in the institutions now being created.

There were two main differences over coverage. Malta did not accept the implications of paragraph 13(a) of the declaration, that the new machinery should deal only with the resources of the sea-bed, and instead proposed an ambitious and elaborate treaty which would establish institutions to manage the sea as well. (7) The idea of an "ocean space treaty" was to be revived briefly in the First Committee at Caracas, and attracted some not uninfluential support. (8)

A more serious difference, in its potential for wrecking or protracting the conference, was that over the stages of mining operations with which the Sea-bed Authority was to be concerned. Canada, and the eleven Latin American and Caribbean States, wanted its writ to extend to "all activities relating to production, processing and marketing", a position to which paragraph 4 ("other related activities") and paragraph 9 ("rational management of the area") of the Declaration of Principles lent some support. The USA, in contrast, sought from the start to confine the Authority's jurisdiction to exploration and exploitation, thus denying it any real means of measuring the value of sea-bed production, so long as that is confined to nodules saleable only in their processed form. This cleavage was to endure.

Within these twelve proposals, differences in respect of redistributive potential were rather more striking than those of coverage. At one extreme France would have left it to "voluntarily-accepted obligations" on the part of exploiting states to contribute "an appreciable share" of the relevant taxes to assistance to developing countries, through programmes that could have been international, regional or bilateral. The Authority's only role in this would have been to monitor such programmes and penalise states that fell down on their undertakings. (9) The Soviet and Polish proposals were also decidedly sketchy in respect of redistributive potential, though they did speak of giving special consideration to the needs of developing countries and indicated that this would be done through some form of international organisation; an advance, this, on their position with respect to the Declaration of Principles. (10)

At the other extreme, the Latin American/Caribbean proposal would have given the Authority wide scope to favour developing countries, not only in the distribution of revenues, but also in respect of technical and scientific assistance, employment, investment in joint ventures, and the siting of processing plants. (11) Tanzania proposed that members should benefit in inverse ratio to their respective contributions to the UN's annual budget. (12) Malta's treaty would have set a maximum limit on administrative expenses, varying with total revenue, and have earmarked the rest for specific projects, related to ocean use, of which at least 15% were to benefit the land-locked. Malta also included a provision, found in the US, Canadian and British proposals, whereby states would also contribute to the redistributive process from the exploitation of their coastal zones. In the case of the British and American proposals this was to have taken the form of an Intermediate Trusteeship Area. (13) For the international area itself, the American proposals were by far the most elaborate; they made clear that the licence fees, site rentals and payments on production which were to be required of each operator were to be specified in the treaty. (14) The residue, after deducting the Authority's expenses and other costs of facilitating exploitation and pertinent research, was to be devoted to the economic development of its developing members via international organisations to be listed in an Appendix. (15) By contrast, Japan's proposal simply left it to the Authority's Council to decide on the financial obligations of miners and on the method of distributing revenues. (16) Canada's proposals for the use of the Authority's revenues were also remarkable for their optimism as to their likely scale. They were expected to "contribute significantly to the economic advancement of the developing countries" as well as serve a variety of other purposes. (17)

These proposals also differed in the extent to which they would restrict the free operation of the "first-come, first-served" principle, embodied in its purest form in the US proposal, whereby the Authority would be obliged to grant any applicant a licence if it complied with the rules. Against this, there was much sympathy for the consequences land-based producers might suffer from sea-bed competition. The debate arising therefore will be discussed in the next chapter. The Soviet Union also expressed some anxiety over the possibility that sectors used for exploitation might form a belt cutting some states off from the three main oceans. (18) Another of its concerns, shared by Poland, was with equity as between one state and another in the allocation of sites. These countries proposed that each state should have the same number. (19) Japan, in similar but much less extreme vein, wanted to limit the number of sites any one state could get in a year from a given category of minerals. (20) The British and French proposals also addressed this same problem, but in almost diametrically-opposed ways.

The former proposed a complicated grid system, whereby "blocks" would have been gradually opened up to exploitation at fifteen-year intervals, for which any member state could apply if it could show it had a potential operator behind it. (21) The latter purported to dispense entirely with the assignment of exclusive rights where 'mobile equipment' was required - that is for the mining of nodules; thus, by implication legalising "poaching". (22)

141

The foregoing will already have given some indication of the differences among these proposals in respect of the role they assigned to a sea-bed authority (or the like), but to get an adequate conception of the full range of views, one must compare the proposals of the USSR, Poland or France on the one hand, with those of the thirteen Latin American/Caribbean states, on the other. As mentioned earlier, for the USSR and Poland, the very provision for the creation of an international body was a significant concession to the Declaration of Principles, but, as they saw it, its task, at least initially, (23) was to coordinate, to discuss and to recommend. (24) France, similarly, would have made the organisation little more than a registry, not even empowered to adopt its own regulations.

The Latin American/Caribbean proposal at the other extreme would have given the Authority almost total discretion to control all stages of the commercial process of feeding metals from the ocean floor into the international economy; and would have enabled the Authority, via an organ called "The Enterprise", to undertake all these activities itself. This was a seminal proposal which was to influence the course of all subsequent negotiation; and in essence it was supported by Tanzania and by the seven land-locked and geographically-disadvantaged states, even though four were European and one (Hungary) socialist.

The remaining five proposals (those of Canada; Japan; Malta; the UK; and the USA) fell somewhere between these extremes. Canada's was perhaps the most hospitable to the possibility of institutional growth, to the point of contemplating that the Authority might some day, conceivably, engage in exploitation. Malta, Japan, and the UK would have allowed the Authority to have pursued a resources policy regulating the pace of production. In the US proposal, which made a very clear distinction between the role of the Authority, and the role of states (and other actors such as companies), the Authority would not have had this power; but the powers it _would_ have had, though circumscribed, would have been substantial. It would adopt rules in fourteen listed fields, supervise compliance with them, and deal directly with applicants, and would have had the capacity to sue and be sued.

It cannot be said that either the First Sub-Committee of the Sea-bed Committee, or its reincarnation as the First Committee of UNCLOS III, had by the end of the Caracas session in 1974 came very near to resolving the wide differences in views as to the system of exploitation which an analysis of these proposals of 1970 and 1971 reveals. In spite of earlier reports of the chairman of the committee's Working Group, Christopher Pinto, that it was "on the threshold of real negotiation and a possible breakthrough" (UNCLOS III, Records, Vol.II 1975, p.45), Caracas ended with the appearance of polarisation. Yet that protracted session offered some clues as to what final

142

compromise, if any, might emerge.

After Caracas it was fairly clear that any agreed system would be more centralised than the extreme laissez-faire of the 1970 proposals of France; would contain some redistributive potential in the form either of financial obligations of contractors, or of profits arising from direct exploitation by the Authority itself; and would cover only the sea-bed and its resources, and not 'ocean space' in general.

The last of these issues was effectively settled after a brief but dramatic flurry in the First Committee, following a plea by Elizabeth Borgese, representing an NGO, the International Ocean Institute, effectively endorsing the philosophy behind Malta's 1971 Draft Treaty (UNCLOS III, Records, Vol. II 1975, pp.30-45). The USA and the USSR joined in insisting that it fell outside the committee's mandate, and though a large number of developing countries, led by Mexico, reaffirmed their right to talk about it, little was heard of it in the conference's proceedings thereafter.

The chief issue that absorbed the First Committee at Caracas, apart from consideration of the economic consequences of sea-bed mining, which will be treated in the next chapter, was that of the "basic conditions of exploitation", into which the protracted pre-Caracas debate on "who should exploit the Area" was transmuted. This debate served to formulate the opposing lines of battle. The Group of 77 had by now followed the Latin American lead on the "who shall exploit" question, (as well as on that of limits to coastal state jurisdiction!) and endorsed a formulation which essentially said that all exploitation and related activities should be done by the Authority unless it decided otherwise. Mention was made of service contracts and the delegation of tasks to 'juridical and natural persons' but only at the Authority's discretion (UNCLOS III, Records, Vol. II 1975, p.45).(25) Now, when asked to set out what they would see as the "basic conditions of exploration" of the new regime, the 77, after a protracted internal debate, which delayed the First Committee for nearly a week, enunciated a set of provisions which in effect reiterated their commitment to a centralised and all-embracing system. (26) Title to the Area and its resources were to be vested in the Authority; the Authority was to determine which part or parts of the Area were to be open to exploitation; it had the option of concluding contracts with "any person, natural or juridical", covering every stage of - operations from scientific research - seen here as incapable of being "innocent" - to transportation, processing and marketing; it would establish procedures and prescribe qualifications of applicants for contracts, and select from among them "on a competitive basis", though with recognition of "the need for the widest possible direct participation of developing countries, particularly the

143

land-locked among them; and that selection was to be final and definitive.

The Authority would also have been able, at its discretion, to enter into joint ventures or any other form of association; and to insist that its partner, either to something of this kind or to a contract, provide the necessary funds, materials, equipment, skill and know-how, depositing a guarantee, transferring to it, on a continuous basis, technology, know-how, and providing training for personnel from developing countries, and employment 'to the maximum extent possible' of their nationals where qualified. It could at any time take such measures as might be necessary to regulate production. In the case of contracts, all responsibility, liability or risk arising out of the conduct of operations was to fall on the contractor. In return for all of this, the contractor was assured of security of tenure, provided he did not violate the provisions of the convention or the Authority's rules and regulations; but even this right was subject to the overall discretion of the Authority to revise, suspend or terminate the contract in case of "radical change of circumstances or force majeure". With the cards so stacked against it, it is difficult to see what would have induced any company, or state, to sign a contract or enter into any other commercial arrangement with the Authority.

The other three proposals, those of the United States (L6), the E.E.C. eight (L8) and Japan (L9), scarcely differed among themselves in more than detail. They agreed in limiting the Authority's jurisdiction to cover only the exploration and exploitation stages; in allowing for direct applications by commercial entities as well as by states or state enterprises; in precluding, by implication, any possibility of the Authority undertaking exploitation directly; and in making these Basic Conditions of Exploitation, including whatever financial obligations might ultimately be agreed, an integral part of the convention. (27) Virtually no provision was made to interfere with economic forces in the allocation of sites. (28)

The British plan for a grid system, with a phased allocation of blocks for which any state could bid, was formally abandoned. Instead, under all three systems now proposed, the pace of sea-bed exploitation would be set, essentially, by the pace at which companies (and states) having the necessary technology applied for, and exploited, sites. The Authority was given no guiding reins, to preserve most, or even some, of the richest areas for future generations. The EEC proposal contained an empty requirement that applicants should indicate how they would ensure the participation and training of nationals of countries 'without sea-bed exploration and exploitation capability'. The Authority was not empowered to specify, even in this respect, what counted as a satisfactory answer.

It was obliged to award the contract if there was no completing application.

Thereafter, the central task of the First Committee at UNCLOS III was to bridge this huge gap between developed and developing states. The successive texts that emerged, from 1975 onwards, followed no very logical sequence, and, as we have seen, not infrequently provoked polarisation; but there were, faintly discernible beneath the surface of negotiations, some indications of what an eventual compromise would look like. The key to this compromise was provision for what came to be called a "mixed", "dual" or "parallel" system. In essence what this meant was that duly qualified applicants for a site, whether states or companies, were to be assured of a contract, on terms enshrined (for the most part) in the convention itself; but that convention would also permit the Authority, through its organ "The Enterprise" (as in the Latin American paper of 1971) to mine the sea-bed (either directly, or in conjunction with developing countries), and make whatever arrangements with state or private enterprises it saw fit in order to do so; and would make sure that the Enterprise had, in practice, the capacity to do these things.

The idea of dual system (and not just a system in which the Authority controlled everything but could bring in other entities if it wished) was not new. Traces of it were detectable in the papers submitted by Canada, and Afghanistan and others, in 1971. At Caracas itself, several developed countries, including Canada, Australia, and Sweden, advocated that the Authority be given a direct role in exploitation; and certain developing countries, among them Jamaica, sought to impose certain constraints on the Authority by requiring that "all activities of exploration" and exploitation should be conducted pursuant to regulations promulgated by the Authority and listing the categories of subjects on which the Authority would promulgate regulations (A/Conf. 62/C.1 SR.9). Interestingly and significantly, one of the states which did not, at Caracas, rule out the possibility of the Authority itself engaging in mining, was a member of the socialist bloc, Bulgaria, whose ambassador, Alexander Yankov, held in high regard in the conference as chairman of its Third Committee, said in his statement to the plenary that

"The international sea-bed authority should be empowered to exercise regulatory and licensing functions and, where appropriate, enter into contractual arrangements with States or undertake exploration and exploitation activities if that was feasible and profitable" (UNCLOS III, Records, Vol. II 1975, p.45).

At the conference's third session in Geneva, in the spring of 1975, both the Soviet Union and the United States indicated their willingness to accept a form of dual system in which the Authority would have a role in exploitation. The former did so publicly, in a working document of 21 March 1975 (A/Conf. 62/C.1./L12) which, like the four presented at Caracas (L6, L7, L8 and L9) constituted a response to the invitation to each main country or group of countries to formulate its own 'basic conditions of exploitation'. By providing that the "International Sea-bed Organisation" might "reserve, for evaluation and exploitation with its own means and resources, not more than ..... per cent of the total area of the sea-bed which is open for evaluation and exploitation by States Parties under contract" (Article 5.2), it became the first developed country to table conditions of exploitation which assigned the organisation a direct role. To be in the van in accommodating the Group of 77 marked a big change in Soviet policy since 1970, when it had abstained on the Declaration of Principles, though the seeds of change could perhaps be seen in the Bulgarian statement at Caracas, quoted above, and in the general support given by socialist states at that session to the concept of the common heritage of mankind. The Soviet proposal remained, however, at variance with the views of Western developed states, in that it envisaged (Article 9) that the area that did not come under the direct control of the Authority should be allocated only to states, and that each state would be assigned the same number of contracts in respect of each category of minerals.

Meanwhile the United States had privately moved even further towards the Group of 77 in a proposal referred to without attribution by Christopher Pinto at a meeting of the First Committee at Geneva, but not made public until a speech by Henry Kissinger in the following April, (29) whereby, (in Pinto's words)

"an applicant for a contract for exploitation activities would be required to propose to the Authority two alternative areas of equivalent commercial interest for the conduct of operations under the contract. The Authority would then have the right to select one of these two areas for itself for exploitation virtually at its own discretion ... That was viewed by some as a useful and even necessary device which would ensure access to the resources of the area by qualified applicants only; and at the same time help the Authority to accumulate commercial data which it would otherwise find difficult or very expensive to acquire. The two-area system was viewed as part of an overall agreement that would also cover matters such as certainty with regard to the opening of areas for continued exploitation and non-discriminatory treatment of applicants".(30)

Generous though this was, in ensuring that at least 50% of the best sites were either exploited by the Authority itself, or "banked" for its future use, it was, like the Soviet proposal, difficult to reconcile with the demand of the Group of 77 that the Authority should have control over the rate of production. Both meant that, while the Authority would have control of its own segment of the Area, it would have no control over the number of contracts granted in the "private"or "state" sector since these would be automatic if the applicant fulfilled the necessary conditions.

The site-banking scheme, as it came to be called, was, then, an important move towards compromise, but it did not go far enough. Before the Republican Administration left office at the end of 1976, its Secretary of State, Henry Kissinger, made three more concessions, in principle, in an effort to convince the developing countries that a dual system could be conducive to their interests. The first came in the course of the same speech at which support for "site-banking" was first announced, an eloquent address to three private organisations in New York in April 1976, while the conference's fourth session was meeting in the UN's headquarters in that city. To allay the fears of the land-based producers, he proposed a "temporary limitation" on production of sea-bed minerals, so as not to exceed the projected growth segment of the nickel market, estimated at about 6% per year. The significance of this concession will be discussed in chapter 8; all that needs to be noted here is that it did not add to the Authority's discretion, since the limitation would be embodied in the treaty, but it did offer grounds for reassurance for those who feared that, even if the private sector was limited to half of the Area, it could not be prevented, given a technological breakthrough, from engaging in a headlong expansion, in utter disregard of environmental, political, and above all economic, consequences.

The other two concessions came during the fifth session (the second in New York in 1976), and were phrased in quite vague terms. The first concerned the financing of the "Enterprise" and indicated that the United States, which could be expected to be a major source of any finance, was willing to ensure, in company with others, that that body was in a position to start exploitation concurrently with the contractors in the private sector. The second consisted in the bare announcement that the United States was willing to agree to "periodic review" of the system. How often such a review could be held, and what action could be taken as a result of it, were not specified; but the words hinted at the possibility that the United States would entertain a system with some flexibility; one that was not tied forever to the conditions specified in the treaty. Some even saw this as an acknowledgement of the temporary character of the dual system itself, an opportunity for the Authority, after due deliberation but without formal amendment of the treaty, to replace it with a unitary system of the kind proposed by the Group of 77 at Caracas, in which, if it did not itself necessarily conduct all future exploitation of the area, it could regulate it at its discretion.

The first of these two speeches of Kissinger also mentioned another topic on which the Group of 77 set much store, that of the transfer of technology. He spoke of establishing incentives for private companies "to participate in agreements to share technology and train personnel from developing countries".

He did not suggest, as many of the Group would have wanted, that applicants for a contract be required to transfer their technology to the Authority. For that reason, his proposal cannot be regarded as a concession; but it is a further area in which concessions were possible without negating the principle of the "dual" system.

Thus the elements of compromise in a "mixed" system of exploitation were as follows. The Authority would be permitted to exploit the Area directly through the Enterprise; it would acquire, through site-banking, sites of equal value to those mined in the "private" sector; it would be guaranteed the loans to enable it to mine such sites; it would be helped to acquire, perhaps even assured of the opportunity of buying, at independently-assessed prices, the necessary technology; and it would be empowered, in certain strictly-defined fields, to issue and enforce rules applicable to all operations. In return, so long as the aggregate production did not exceed, in any one year, a (possibly temporary) production limit to be derived by methods to be prescribed in the treaty, qualified applicants were to be assured of their right to conclude contracts with the Authority, and have them fulfilled, on terms, including financial terms, which were also to be specified in the treaty. To reinforce that assurance, any state, contractor, or discontented applicant for a contract would be able to challenge, before an independent tribunal, any decision of the Authority affecting it, him or her, whose decision as to whether or not it was in conformity with the convention was to be binding.

Dispute settlement was thus an essential part of the bargain. It was seen by developed states in general, and the United States in particular, as the way in which they could ensure that no applicant was refused a contract without good cause, and that a contract, once granted, was fulfilled by the Authority.

It was a bargain, essentially, between the capitalist states of the West, led by the United States, and the Group of 77. The Soviet Union was also prepared to contemplate a compromise of this kind, with appropriate changes. Like the developed West, it mistrusted the discretion of international institutions, and sought to protect potential

148

exploiters of the sea-bed from its arbitrary exercise; only in the case of the Soviet Union the potential exploiters were of course not companies but states.

The proposed institutions were distrusted, not because they would be endowed with physical power - there was no question of equipping them with the military or economic means to enforce their writ - but because they would dispense legitimacy. Countries that intended, or whose companies intended, to invest what in the end proved to be more than a billion dollars a time in nodule mining projects were not prepared to see the risks inherent in such investment compounded by the idiosyncracies of a possibly capricious Sea-Bed Authority. They wished so to write the treaty, and so to shape the dispute-settlement mechanisms attached to it, that it was clear, in any given case, whether or not that Authority had acted within the legal powers conferred on it. (31) The implication was that if it was shown that it had not, the Authority would redress the situation, and if it did not, the powerful could intervene and defy it, at least if their own interests were involved, since it would then have lost the legitimacy with which it had originally been invested. Without such strict circumscription of its proper functions, it could be argued, states would be prompted to intervene on their own, or their nationals', behalf on purely subjective impressions of whether the Authority was acting within its rights.

THE QUASI-COMPROMISE OF 1980

Before turning to the quasi-compromise of 1980, embodied in the third revision of the ICNT and greeted with almost universal enthusiasm, it is necessary to remind ourselves that the system of exploitation of the common heritage of mankind that was being negotiated, was always formulated as applying to all mineral resources that might be found in the Area, and not just manganese nodules, although there is no doubt that it was with manganese nodules in mind that most of the provisions were drafted. Article 137 of the convention provides that rights with respect to any of the minerals of the Area could be 'claimed, acquired, or exercised' only in accordance with the convention's provisions, and article 133 indicated that the minerals so referred to covered all that might be derived from

"all solid, liquid or gaseous mineral resources _in situ_ in the Area at or beneath the sea-bed, including polymetallic nodules".

Article 137, and a less elegantly-worded version of Article 133, go back to the Single Negotiating Text of 1975 and do not seem to have been very controversial before 1982 (32) when, as we shall see, the US delegation began to express some concern about the restrictions they might impose upon the exploitation of the non-nodule resources of the Area (US Delegation 1982, pp.20-21). Before that, the only point of contention has been the fact that the ICNT, issued in 1977, had permitted the Council to impose limitations on production of such resources, a change made by Engo to the compromise texts drafted by Evensen, which was quickly reversed at the first part of the seventh session in 1978. Otherwise there was no dispute that, while the rules for non-nodule resources might be different, for instance with respect to size of area, and the the possibility of the same terrain being contracted to two different parties for different types of mining was not to be ruled out, the basic conditions of exploitation, and, with the exception of production limitation, the whole of Part XI of the convention, should apply equally to any resources that the Area might be found to contain.

To translate the parallel system into reality, three sets of questions had to be answered. One concerned assurances to potential miners, whether states or companies. This set of assurances could also be divided into three: on what terms would contracts be offered; on what grounds could an applicant for a contract be refused one; and what guarantee would a contractor have that his contract would be honoured? (33) The second set of questions related to the Enterprise. Would the convention ensure that it would have the practical option of going into business simultaneously, and if so, how, and what constraints, if any, were to be placed on its freedom of action. The third set of questions concerned the permanence of the initial system. What could be changed, and how, and what would happen if there were no agreement on what changes should be made?

The approach to an agreed system of exploitation by this route was an arduous one, and there have been those, among both developed and developing states, who have discerned a more manageable alternative through a unitary system of joint ventures between the Authority and public or private enterprises. During the third session at Geneva, for instance, the Minister in charge of the British delegation, David Ennals, suggested at a press conference that the idea of such ventures offered the best basis of agreement, adding, in answer to a question, that this did not mean that the terms had to be set out in full detail in the convention, and that he thought that they could in large part be left to the Authority to decide: 'it was not sensible to write an inflexible form of contract into the convention'. This was taken up and explored by the working group and an informal paper circulated on how joint ventures might apply to the sea-bed. The idea found expression in a provision of the Informal Single Negotiating Text that the Authority should immediately identify ten economically-viable sites and enter into joint ventures in respect of them (Part I, Article 22). It was resurrected at the end of the deadlocked fifth session at New York in September 1976, when Nigeria circulated a paper which won considerable support among both developed and developing countries. Negotiations under Jens Evensen's chairmanship, however, at the intersessionals of February 1977, and during the first three weeks of the sixth session later that spring, followed a different course, and when, at the latter, Pinto attempted to reopen the question, he had little encouragement from the chair. Even that, though, was not the last to be heard of it. It emerged, as we shall see, in the provisions of the ICNT of 1977 as to what would happen if the review conference resulted in deadlock, which were doubly unacceptable to the West, firstly in that by them, joint ventures would replace only the private (and state) track of the parallel system, and secondly as further examples of the First Committee chairman, Engo's, tampering, as they saw it, with the compromise arrived at by Evensen.

As late as 1979, Elizabeth Borgese, of Malta's International Ocean Institute, and now attached to the Austrian delegation, was pressing the idea of joint ventures as a unitary system by means of seminars and through the somewhat delayed publication of her booklet, "The Enterprises"; and in the New York part of the eighth session the Dutch delegation put forward a modified version of it whereby the Enterprise would have the option of taking a 20% holding in the stock of any applicant in return for making a similar offer in respect of one of its sites.

In spite of such campaigns, however, and of the attempt of the United States, in late 1978, to interest the conference in a simplified treaty, the details of which remain undisclosed, the questions of balance, access, and permanence, which lay at the heart of the parallel system, could not be avoided.

Perhaps surprisingly, in an age in which reliance on it in other aspects of international economic relations is noticeably diminishing, (34) the principle of sanctity of contract proved the least controversial element of the negotiations over access. As we have seen, the Group of 77 at Caracas wanted to allow for exceptions to it, and the SNT of 1975 specified circumstances, including "force majeure", that might justify the Authority in revising a contract, (35) but in the RSNT these qualifications were taken out. The chief remaining power the Authority enjoyed to suspend a contractor's rights, by the issue of a stop-work order to prevent serious harm to the marine environment, (36) did not represent a concession by the United States; it was part of that country's 1970 proposal discussed earlier. Otherwise, a contract could be terminated only if, in spite of warnings by the Authority, the contractor had conducted his activities in such a way as to result in serious persistent and wilful violations of the terms of the contract, the relevant provisions of the convention and the rules and regulations and procedures of the Authority or failed to comply with a final binding decision of an applicable body. (37)

Sanctity of contract was not, of course, enough for potential mining states. They, or at least the USA, wanted to whittle away the Authority's discretion to the point where it could refuse an applicant only on certain specified grounds, and where the terms of the contracts it could offer were virtually dictated by the convention itself. As Leigh Ratiner put it in 1974

"We need to establish conditions of exploitation which by their terms provide for sufficient security and stability and guarantee of non-discrimination that when we go to our Senate, hopefully within a year from now, we can testify that the United States is protected ... that the conditions of exploitation in our judgement will cause private enterprise to want to produce minerals and .. that, if they elect to ... they will have an opportunity to return those minerals to the United States where they will be consumed" (Ratiner 1974).

Nevertheless, as the conference proceeded, it became clear that the convention, for all its bulk, could not answer in advance every question that could be asked about how exploitation would proceed. The Authority (or the Preparatory Commission) would somehow have to be left with the capacity to adopt "rules, regulations and procedures" and to take a wide range of other decisions, implementing the basic principles of the convention. The structure and decision-making rules of the Authority will be discussed more fully later, in Chapter 9, but it is necessary to anticipate this discussion somewhat here.

Although voting rules in the Council were not settled until the summer of 1980, it was generally clear that the developing countries would have a preponderance of votes even here, though not as marked as in the Assembly in which all members would be represented with equal weight. The fear then was that even the Council might adopt rules, regulations, and procedures which, if they did not, in effect, annul the assurance of access otherwise afforded by the convention, might attach to it what the mining states saw as unforeseen, unpalatable and unnecessary burdens and constraints. For long, then, negotiation focussed on providing the West with an adequate brake on the Authority's capacity to adopt such rules. The accord of 1980, however, which furnished every member of the Council with a veto over certain decisions, including the adoption of such rules, eliminated the possibility that rules etc., obnoxious to mining states might be introduced, but left open the equally distasteful possibility that none would be adopted at all and that therefore no contracts could be entered into. On the other hand, once such rules had been adopted, it would not be easy to alter them, and certainly not over the heads of the mining states. Thus if anxieties over access were to be fully allayed, before ratification, these rules would have to have been previously elaborated, which, if the mining states were to have a direct part in this elaboration, would have meant by the Preparatory Commission, of which, it was later decided, those that signed the convention would be members and those that signed only the conference's Final Act would be observers.

There was no major controversy as to the fields in which such rules would have to be drawn up, though there were some differences of view, and some changes, particularly as between the SNT of 1975 and the RSNT of 1976. The developed countries at Caracas had proposed to write into the convention, in the annex setting out the 'Basic Conditions of Exploitation', the size of area and length of time for which contracts were to be issued. From 1976 onwards, there was agreement that it should be left to the Authority to determine these points, though with the assistance of verbal guidelines in the "Basic Conditions". On this the precedent set by the SNT was accepted. In that document, a large number of topics had been left for regulation by the Authority; the US, in 1975, sought to reduce the list, with some though not total success, and the version appearing in the RSNT was retained, with only superficial changes in subsequent texts including the convention itself.

These rules were to be uniformly applied. It could be inferred that they could be repealed or amended by a process similar to their enactment. No distinction is made between rules that might apply only to new contracts - eg those concerned with size and duration of contracts - and those that, presumably, would govern existing contracts - eg those relating to mining standards, operational safety, resource conservation, and protection of the marine environment. The consensus requirement for the adoption of such rules, agreed in 1980, ensured, by its very stringency, that this difference would not be of crucial importance; it also severely diminished the Authority's capacity to adjust its rules promptly as new knowledge became available.

The clear assurance that the Authority would be able to refuse an applicant a contract only on specific and highly-circumscribed grounds took a long time to emerge. The SNT was highly ambivalent. It conferred wide discretion on the Authority to negotiate with applicants, and to assess the benefits each offered to the Authority: while seemingly denying the Authority the right to reject a qualified applicant who had met all the conditions, and affording any such disappointed applicants a right of appeal. The RSNT, incorporating some of the amendments proposed by the USA, narrowed the extent of this ambivalence without eliminating it (a quid pro quo for US acceptance of the principle of production limitation). At the fifth session, when the Group of 77 had rejected the RSNT, the USA and Britain explicitly dissociated themselves from a remark by the co-chairman of the Committee I workshop that

"It is doubtful that any delegation supports an

153

automatic assurance of access, since there seems to
be general agreement that the Authority will
presumably have some degree of discretion in
applying the relevant provisions of Annex I" .(38)

Discretion was just what Leigh Ratiner sought to
eliminate. It was what Elliot Richardson sought to
eliminate, too, though it took some time for him to
accomplish. The Evensen drafts which he was prepared to
accept as a basis for negotiation contained some slippery
phrases about transfer of technology and the "degree of
participation in the project by developing countries". That
is, the Authority was to "negotiate" with the applicant on
these matters with presumably, the right to reject an
application deemed unsatisfactory in those respects (and
also, explicitly, to give privileges to the more generous).

Engo changed this, in the ICNT, by making the transfer of
technology (on fair and reasonable terms, subject to binding
arbitration in case of dispute) a condition on which the
Authority could insist, and also by giving the Authority
apparent power to require that applicants be willing to
engage in joint ventures. These were two of the changes to
which Richardson reacted so sharply in his response to the
ICNT and it was with some assurance that they would be
reconsidered, in Frank Njenga's Negotiating Group 1, that
the US Delegation, following its review of the methods and
prospects of UNCLOS III, agreed to participate in the
seventh session. At the end of the Geneva part of this
session, Richardson was sufficiently satisfied with the
progress made to suggest, in his delegation report, that
recognition had at last been given to the principle of
untrammelled access of qualified applicants. A year later,
in the spring of 1979, it was embodied in the ICNT, which
included a provision, Article 6.3 of Annex II (later to
become Annex III) listing the grounds on which the Authority
could refuse to conclude a contract with a qualified
applicant willing to agree to the convention's terms and the
Authority's rules.

In that text, these grounds were four. The first, in no
way controversial, was if it coincided with, or overlapped,
a site for which a previously approved, or pending,
application had been made. (Refusal in such a case follows
from the notion of exclusiveness of rights). A second
ground was if the area fell within a zone which the Council
of the Authority had disapproved for exploitation because of
"the risk of serious harm to a unique environment." This too
had been a non-controversial feature of the ICNT,
strengthened by verbal amendments in Revisions 1 and 2,
which made the grounds on which an area could be so
disapproved less exacting. When the voting system for the
Council was agreed in 1980, this type of decision fell
within the category requiring a three-quarters majority,
thus leaving that organ some discretion; but it is required
to act, in such cases, only on "substantial evidence".

154

A third ground for disapproval of a plan of work was, in the first revision of the ICNT, if production permitted under it would bring total production above the specified production limit. From the third revision in September 1980, it was provided that the production limit was to be administered through "production authorizations" under Article 7 of this Annex; but in substance, as we shall see in the next chapter, the constraint remained. The fourth ground of rejection was that it gave any one state (or entities supported by it) too large a share of the total area available for exploitation of sites. At Caracas, the USA had opposed any such limit on the number of applications any one state could sponsor. The EEC countries, while seeking to limit the number of sites any one entity could control, had otherwise supported the American position. Strict rationing of sites by states was, as we have seen, a feature of the Soviet proposal of 1975 (as of previous socialist state proposals); and, as the production limit began to look as if it might bite, the French and British showed indications of support for such rationing, in order to ensure that US companies did not acquire the right to take all the permitted production. From 1975 onward, though, the USA showed itself willing to concede some limits of this kind, and in 1979 raised no objection to limiting each state to not more than 3% of the total eligible area, and not more than three sites within any one circular area of 400,000 sq.km. In the second revision, at the instance of France, these limits were made more stringent: 2% of the total area, and 30% of any circle of 400,000 sq.km drawn from the centre of each relevant area (ie. each of the two offered by the applicant from which this Authority would choose one). In return, the US at least secured that the limit should be based on sites actually held, neglecting any abandoned or exhausted. Moreover, the Authority could waive it if it saw no danger of monopoly in a given application.

Since under it a contract could not be disapproved except on one of the foregoing grounds, the second revision of the ICNT in April 1980 seemed to give mining states the "untrammelled access" they sought; this impression was reinforced by the reinstatement in the third revision of a provision to the effect that after a specified period (sixty days) a plan of work which had satisfied the Technical (later the Legal and Technical) Commission would be deemed approved by the Council in the absence of a vote to disapprove it. The voting rules agreed in the summer of 1980 required that such a decision to disapprove be taken by consensus, excluding the sponsoring state or states. Thus if the Commission could be relied on simply to apply the criteria of Article 6.3 of Annex III, a plan of work could not be arbitrarily overruled. The second revision briefly disturbed the clarity of this assurance by amending Article 162 2(j) so that any vote on a proposal to approve or disapprove a plan of work could invalidate its automatic approval; a further eccentricity of the First Committee's chairman, Engo, which was corrected in the third revision later that year. In any case, were the Legal and Technical Commission to recommend disapproval, or the Council to vote to that effect, on grounds other than those listed in Annex II Article 6.3, the applicant could seek judicial redress.

The final element of assured access which potential mining states were anxious that the treaty should contain was a clear statement of the terms on which contracts were to be granted. What did the applicant have to offer in return for a site? Distrust of such an entirely novel phenomenon as a sea-bed authority led to the demand that it be given little, if any, room to bargain in this respect, and thus that the extent of a contractor's obligation should be set out in the convention itself. Although some of the proposed obligations proved largely, or entirely, non-contentious, two, in particular, over financial payments and technology transfer, led to some of the conference's most protracted negotiations.

Several of the less contentious obligations merit some mention. By Article 153, contractors would have had to undertake not only to comply with the convention and its annexes and the rules, regulations and procedures of the Authority, but also to accept the Authority's control (as far as might be needed to ensure compliance), including the right to inspect all operations; they would have had to fulfil such performance requirements as the Authority would prescribe - that is they could not just sit on a site without spending any money on developing it (Annex III, Article 17.1(b)(iii) and 17.2(c)); and to draw up programmes for training the personnel of the Authority and of developed states and their participation in all activities in the Area covered by the contract (Annex III, Article 15). More important, and indeed, as has been mentioned earlier, the cornerstone of the parallel system, was the "site-banking" scheme, that is the requirement that each applicant offer the Authority the choice of two sites (or two halves of one site). By Article 8 of Annex III the applicants have to submit, for applications to mine nodules, data relating to mapping, sampling, density of nodules and mineral content, for both parts, and the Authority shall select one within 45 days which then, as soon as the plan of work for the non-reserved part is approved. Reserved areas would be exploited either by the Enterprise or by developing countries or by some combination of both. This obligation on contractors was never seriously disputed; it tended at the conference to be undervalued.

We come now to the two sets of obligations of contractors on which agreement proved most difficult: financial payments and technology transfer.

The discussion of financial arrangements came late in the
order of business at UNCLOS III. There was an exchange of
views on it, apparently "confined to available experts from
a limited number of countries" (RSNT PART I, p.4) at the
fourth session in the spring of 1976, and two approaches
were set out in a special appendix to the RSNT. The chasm
that divided developed and developing countries at the fifth
session discouraged further work on it, and it was not until
the Evensen discussions in Geneva and New York had improved
the prospects of agreement on the principles of the system
of exploitation that the conference took it up in any depth,
in the middle of its sixth session in 1977. By then
prospects for profits in nodule mining looked considerably
less sanguine than they had done in 1974. (39) The
difficulty of agreement, though, seems to have come as a
surprise to delegates.

The main types of financial obligation envisaged were
production levies (defined as a certain percentage of the
value of production) and profits taxes. In specifying the
level of these, two very recalcitrant issues arose. First,
since nodules themselves were not, immediately, saleable,
the value of output, and the profits thereby generated,
could be measured only after they had been processed. Yet
the processing itself, at any rate if it took place on land,
would necessarily come under the jurisdiction of some state
or other and could not be taxed without imposing on the
operator a risk of double taxation. Indeed, it was argued
that since transport and processing were not activities of
"the Area", it was not in any case appropriate to tax them.
It was therefore necessary to attribute to the mining stage
a certain share of the value of the processed nodules, and
in the absence of a market for nodules, the proportion to be
so attributed was necessarily hotly disputed. Secondly, the
capitalist states preferred profits taxes to production
levies, because the obligations would fall on the operators
only after they had recovered their costs: whereas the
Soviet Union preferred production levies since there was no
recognisable measure of profit in its enterprises, and
indeed an attempt to audit one of them would hardly have
been able to stop short of auditing the entire government of
the Soviet Union. Thus a solution had to be found that
offered two systems of payments, but nevertheless ensured
that the burden imposed by each, and thus the gain of the
Authority, was equivalent.

At the sixth session a small group under the chairmanship
of John Bailey of Australia was set up to deal with these
problems and on June 6th it received two contrasting
proposals from India and the United States.

The Indian plan provided for both a production levy and an profits tax. The levy was to take two forms, a flat-rate 'royalty' of $5 per ton of nodules actually mined, and a 'tax' of 20% of the revenue from the sale of processed metals derived from them. The profits tax was to be 60% of any net proceeds accruing after the contractor's return had exceeded 200% of his investment. The plan would also have imposed two further charges on the contractors; a small application fee ($100,000), and a "charge to mine" of $1 per ton of nodules authorised to be mined over the duration of the contract, payable within a month of the award of a contract. (40)

The United States proposal made the two main forms of payment alternatives, combined in each case with an application fee. The fee itself could have been higher under the American proposal than under the Indian one, that is up to $500,000. A contractor choosing the profit-based alternative would make payments depending on the "net proceeds" of the mining stage of the operation in each year, and the overall rate of return on investment up to that point. There would be no payments while the rate of return, overall, was zero or negative. Once it became positive, the contractor would pay 15% of such net proceeds as did not bring the rate of return above 15%, 25% of any remainder until the rate of return exceeded 25%, and 50% of any such excess. The other mode of payment consisted of a fixed percentage of the imputed value of the unprocessed nodules, which was to be fixed, apparently permanently, at one-fifth of the "gross fair market value" of the minerals derived from them. Of this "imputed value" the contractor would pay 10% in the first ten years and 50% thereafter (ie 2% and 10% of the value of the processed metals) with an option of switching to the profit-sharing scheme at that point.

Thus while, in the United States plan, the "profit sharing" arrangement was an alternative to the production charges, in the Indian plan it was an addition to it; and, if the Americans were correct in imputing the value of the nodules at 20% of the value of the metals derived from them, the rate proposed by India would have been 100% of that value, plus at least $6 a ton in the form of "charge to mine" and "royalty"; in other words, under it any mining operation would have made a loss of at least $6 for every ton of nodules mined, even before considering the investment and operating costs of the mining itself. When, five years later, the government of India itself sought, and obtained from the conference recognition a a pioneer investor, it was no doubt relieved that these proposals had not been accepted.

Quite apart from the _level_ of payments, the Americans
objected to the Indian plan because of the timing and nature
of the payments due. Under it, the contractor paid heavily
from the start (with the "charge to mine") and his payments
further increased sharply when production began. The
Americans objected to this as "front-end loading"; that is
burdens incurred from the start, and unrelated to success.
They would have preferred a scheme with high rates of
payment if and when an operation proved profitable, and low
or zero rates if it did not, to one with such fixed charges
payable at the beginning as would, after future payments had
been appropriately discounted, average out the same for an
average project over its whole span.

The ICNT included what was described as "a preliminary
draft" of the financial terms of contracts (Annex II,
Article 7). No figures were included, but all four kinds of
charges suggested in the Indian proposal were provided for.
When, as has been described earlier, the process of
negotiating the "hard core issues" of Committees I and II
were restructured so as to be assigned to seven negotiating
groups each with different chairmen, Negotiating Group 2,
under the chairmanship of Tommy Koh of Singapore, later to
succeed Amerasinghe as President of UNCLOS III, was assigned
the task of reconciling these diverse proposals (as well as
that of providing for the financing of the Enterprise, which
the Group of 77 then saw as a closely related question).

Koh began by attempting to reach agreement on the _types_ of
charges to be levied. He suggested retaining the "annual
fixed fee" (what India had called the charge to mine), but
accepted the American contention that contractors would have
to be offered a choice between two systems of payments.

A variety of states then made proposals as to the levels
at which these payments should be set. The contention of
the United States and others that only 20% of the value of
finished metals produced from nodules could be attributed to
the mining stage was supported by other mining states but
evoked some scepticism elsewhere. Two countries made
proposals in terms of production charges only: the USSR
suggested that these be set at 7.5% of the market value of
the processed metals; Norway at 8% for the first five
years, and 16% thereafter. There were five other proposals
including a further one from Norway, for a "mixed" system of
payments, consisting of a smaller production charge and a
percentage of net proceeds. India, at one extreme suggested
that these should be, respectively 10% and 50%; Norway 3%,
and 25% of net proceeds in the first ten years, and 5% and
40% in the second ten; the USA, a 2% production charge and
6%, 12% and 15% of net proceeds according as the rate of
return was less than 7%, 7% to 20%, or more than 20%. Other
potential mining states were less generous. Japan and the
EEC proposed a much lower production charge (0.75%) and
rather lower rates of profit tax (Japan, 5% of net proceeds
rising to 10% for the last ten years, the EEC, a rate that

159

varied, according to the contractor's rate of return, from 2% to 12%).

The wide difference among these proposals in their implications for what the Authority would receive was illustrated by the MIT study mentioned earlier. Assuming a twenty-year mining project with development costs of $559 million, annual operating costs of $100 million, and gross proceeds of $258 million a year, the take of the Authority is shown in Table 7.1.

Remarkably, even the less generous of the "production charge only" tariffs would have given the Authority more than any proposals for a "combined" system emanating from potential mining states. On these assumptions, the Authority would have gained nothing from 'profit sharing' to compensate it for its three disadvantages: uncertainty, a later distribution of payments, and a greater administrative task in assessing profits. Moreover in all these cases the redistributive potential is less - in all except the Indian proposal much less - than that envisaged in the Secretary-General's study of 1974 where each three million ton a year operation was expected to generate between 76 and 118 million dollars a year. (42)

Table 7.1
      The Authority's Gain from the "Private" Sectors
      under Proposals made to NG1 at Geneva, 1978.(41)

Proposing country or group of countries	Type of System	Total income accruing to Authority from a 20 year project. ($ million)
The USSR	Production Charge Only	390
Norway	Production Charge Only	728
India	Combined	1600
Norway	Combined	1050
USA	Combined	335
Japan	Combined	240
The EEC	Combined	75

Koh's first concrete proposals, issued during the second (New York) part of the seventh session, were nearer to the upper end of the scale, higher than those of any national proposal except India's. (43) In subsequent revisions he gradually moved closer to the position of the mining states, and his proposals in the summer of 1979 were included in the second (and third) revision of the ICNT, though France and Belgium claimed that they were too high and India that they were too low. The chief difference between these and earlier proposals lay in the fact that they provided a basis for calculating the contractor's net proceeds attributable to the mining stage, while specifying that in a three metal operation, these should never be deemed to fall below 25% of

160

the contractor's total proceeds. A simple indication of how the burden of charges had been reduced in comparison with his original proposals is given by the production charges imposed on those opting for such charges only. Instead of 7.5%, 10% and 14% of the value of the processed metals, they became 5% in years 1-10 and 12% thereafter. In the mixed system, they became 2% until the contractor had fully recovered his development costs and in any subsequent years in which the return on investment fell below 15%, and 4% otherwise; plus a sliding scale of percentages of attributed net proceeds, similar to income tax ranges, from 35% to 50% until development costs had been fully recovered, and from 40% to 70% when they had been. Thus what emerged as the Authority's take was a smaller proportion of a smaller total, since the new formula would almost certainly attribute a smaller proportion of net proceeds to the mining stage than Koh's earlier proposals had assumed. Moreover, the "charge to mine" of $1 million per year was to be merged in the production charge as soon as the latter began to exceed it.

This scaling down of the payments contractors had to make was part of a package negotiated within Negotiating Group 2, which was also concerned with the financing of the Enterprise and the Authority. The idea that these payments, if high enough, would give the Enterprise the finance it needed to go into business, had been a major obstacle to agreement. As an alternative way of securing this end, it had been suggested, as early as the sixth session, that all states parties to the convention contribute enough to the Enterprise to enable it to exploit one site, if it so chose, at the outset. By amplifying this offer, and convincing developing country delegates that it was serious, the developed states, particularly the USA, were able to dispel much of the opposition to reducing the charges contractors were to pay.

The agreement, embodied in Article 13 of Annex III, ran to nine pages. The complications, and the attempt to provide elaborate and precise definitions of key terms, were a reflection of the uncertainties of the field, and the widely varying range of knowledge among delegates, coupled with an unwillingness to accept on trust the assertions of those whose connections with ocean mining made them well informed but suspect. A straight "production charge" would have been the simplest to negotiate and administer, but both the West and, on the whole, the developing countries, for whom "production charges" smacked of "royalties" associated with "privileged" foreign investment, preferred profits taxes. Still, the negotiating group had coped with this aspect of its mandate well, if lengthily; so much so that it was not among the features of the system of exploitation that the United States sought to change in 1982 after the Reagan review.

The most difficult element among the terms on which contracts could be signed, and indeed, it might plausibly be claimed, in the whole business of negotiating a system of exploitation, was transfer of technology. Among many developing countries, this was the chief gain to be looked for from such a system, more important, even, than the redistribution of resources. Yet if it were to be left to the goodwill or economic interest of exploiting companies, and states, technology transferred might be inadequate or excessively costly. At Caracas, the US delegate, John Stevenson, had pledged full support to attempts to endow the authority with "procedures and mechanisms" to ensure the execution of commitments for such transfers, though at this time any parallel system was ruled out and the recipients would have been the developing countries themselves (Stevenson, First Committee 1974). As the parallel system came to be accepted, technology transfer came to refer to transfer to the Enterprise.

Earlier texts had made this a matter for negotiation between the Authority and an applicant for a site. This approach had been followed in the Evensen proposals of 1977 and the ICNT; but in the latter, there was also a provision requiring an applicant to undertake that, should the Authority so request, he would make available to it under licence any technology he used, on "fair and reasonable terms", to be determined, in the event of disagreement, by arbitration. Failure to fulfil this undertaking could lead to suspension or termination of the contract. The developed countries tended to object to both these approaches. The earlier one, because it was vague, gave scope for the Authority to use its discretion and thus threw doubt on whether an applicant would get a contract at all, and if so on what terms. The second, if it had replaced the first, would have removed that doubt; the applicant would have received a contract if it had given the undertaking; but that undertaking was seen as impossibly broad. It was claimed that the applicant did not even have the legal power to transfer some of the technology it used, which implied that it was supplied by some legally independent firm which was prepared to make it available to the applicant but not to the Enterprise. Certainly it was one of the features of ICNT that Richardson found unacceptable and insisted must be renegotiated.

In the next three years the obligation was gradually whittled down, though it was also, in one respect, extended, since any technology liable to be transferred to the Enterprise, would by what became known as "the Brazil clause" also be required to be transferred to developing countries applying to mine "reserved" sites. Richardson claimed that as a result of the spring 1978 revisions "transfer of technology is no longer a precondition of obtaining a contract to mine", and the clause that seemed to suggest it was (5(d)(iii) of what was then Annex II) was dropped from the ICNT's first revision.

Applicants were asked to describe the technology they proposed to use, to undertake to transfer that part of it which they were entitled to transfer, and to seek a written assurance from the owner(s) of any remainder to make it available to the Enterprise. If the owner were to default on this assurance, this would put him in the Authority's bad books, but not necessarily lead to action against the contractor unless a link between them could be demonstrated. A further limitation was that the Enterprise was to invoke this only if it could not obtain similar or equivalent technology on reasonable terms in the open market (Article 5.1(c)); and the second revision of the ICNT, in 1980, restricted the obligation to contracts made before the Enterprise had started commercial production, and allowed it to be invoked for only ten years after that date. In the third revision this later date was also to apply to the inclusion of the obligation in contracts. These limitations imply that soon after the Enterprise has started in business it will have no further need for technology transfer. Yet it would be surprising if there are not further technological advances in nodule mining after this ten-year period has expired, which the Enterprise might want to acquire; and if, subsequent to this, other forms of mining in the Area prove profitable, the Enterprise will have to rely on means other than this clause to acquire the technology necessary to engage in them. The Brazil clause was also narrowed in its application, so that it could not be invoked where the Enterprise had not also invoked the right of compulsory purchase, and the technology so acquired could not be transferred to a third state or its nationals.

This result rather fell between two stools. It allowed compulsory technology transfer to play only a limited role in enabling the Enterprise to compete with private and state mining ventures, while arousing the sustained hostility of those who opposed any such obligation on principle. It did not destroy the principle of "assured access"; but by the requirement not to use any technology outside their disposal without furnishing a written assurance from its owner that it could be made available to the Enterprise, it probably added to the obstacles applicants had in meeting the conditions for the award of a contract, to a greater extent that it benefitted the Enterprise. In 1980, although Richardson insisted that the USA remained committed to the deletion of Article 5 paragraph 3(e) of Annex III (the "Brazil Clause") (Richardson 1980, p.4), he did not seem to regard the obligations to transfer technology to the Authority as impairing the acceptability of the package.

In looking at the quasi-compromise of 1980 we have so far concentrated on the rights of access of applicants and the conditions to which they might be subjected. The other side of the parallel system, the Enterprise, has been dealt with only in passing. It was part of the deal of 1980 - and the reason why it was broadly acceptable to the Group of 77 - that the Enterprise should be not only entitled to mine, but equipped to do so. This meant furnishing it with four kinds of assistance; sites, finance, technology, and given the Authority's production policy, production authorisations.

We have seen how, from 1975 onwards, there was general agreement that the parallel system should automatically supply the Enterprise with promising sites, by requiring each applicant to offer the Authority a choice of two. The offer was to be accompanied, in the case of applicants for nodule mining, by data about their terrain and the density an composition of the nodules in them. (44) The Enterprise would thus acquire a "bank" of "reserved" sites, which it could itself exploit, or retain, except that if there were an application to exploit a "reserved" site from one or more developing countries, or their companies, the Enterprise would either have to decide to exploit the site itself, or allow the application. The Enterprise could also apply to exploit any other part of the Area (Annex III, Article 3.2).

As we saw earlier in the chapter much of the conference's discussion of the financial obligations of contractors tended to assume that, just as private and state applicants would furnish the Enterprise with sites, so also, through their financial payments, they would be the main source of the Enterprise's finance. Yet the sums required were likely to be so high that this was not a very realistic hope. An alternative had been floated by Kissinger at the fifth session (45) and was elaborated by Richardson in the intersessionals before the sixth session in 1977, namely that all member states voluntarily agree to be "contingently liable" for sufficient debt as would enable the Enterprise to mine its first site. This implied a combination of direct loans and undertakings to guarantee loans obtained from other sources. Only a minority of it was to be found immediately in cash; and, armed with these guarantees, the Enterprise would be expected to attract lending from commercial sources and "international financial institutions"; in Richardson's words,

> "it does seem clear to us that, if we are to have a fair parallel system with assured access for states and state-sponsored contractors on the one side, we must also assure that the Enterprise can, indeed, get into business" (Richardson 1977).

The ICNT incorporated Richardson's suggestion in a revised draft of the Statute of the Enterprise whereby states parties would guarantee debts incurred by the Enterprise in covering the costs of its first site and would advance "as refundable paid-in capital" an as yet undetermined proportion of the liability thus incurred (ICNT Annex III, Para 10(c)(iii)). Parties also undertook to change the constitutions of international financial institutions to enable them to lend to the Enterprise, an awkward commitment, which Richardson was glad to see removed in the texts produced by Negotiating Group in the spring of 1978, which also specified one-third as the proportion of their liability that states would initially advance to the Enterprise." (UNCLOS III, Records, Vol. X 1978, NG2/5, p.57) In the spring of 1979, the first revision of the ICNT included a provision, inserted, according to Richardson, in spite of US objections, which raised the proportion of the Enterprise's capital that states would have to advance at the outset from one-third to one-half, and characterised it as "long-term interest-free loans." Later that same year, when, in return, he had persuaded Negotiating Group 2 to accept a lower-level of payments by contractors, Richardson agreed to this change, and it was retained in the second and subsequent revisions. This revision also allowed for an additional contribution of not more than 15% of the interest-free loan and not more than 10% of the cash advance, to compensate for the shortfall caused by any states not ratifying the convention. This aroused considerable opposition, and in the third revision it was watered down so that the Assembly would examine the extent of any shortfall and, by consensus, adopt measures for dealing with it. Essentially, though, Richardson was successful in persuading delegates to accept that the finance to put the Enterprise in business should, as he had suggested in 1977, come from capital subscriptions and guarantees by States Parties and not from payments by contractors.

A similar approach was attempted to deal with the acquisition by the Enterprise of the technology it needed to mine its first site, though it was not accepted by the Group of 77 as a substitute for the obligations of contractors described earlier. From the first revision onwards, this was supplemented by requiring that states directly engaged in, or sponsoring, mining in the Area, and any others having access to the relevant technology, should consult together so as to ensure that it was made available to the Enterprise "on fair and reasonable terms and conditions" (Annex II Article 5.3). In the second revision this was strengthened by authorising either the Council or the Assembly to convene a group of such states and requiring each State Party to do all it could within its own law to compel firms to sell the Enterprise what it might need (Annex II, Article 5.3). It should be noticed that the technology in question included that needed for nodule processing not just for mining.

The question of production authorisations, which derives from the production limitation on which the Group of 77 has strongly insisted, belongs to the next chapter. Somewhat surprisingly, little thought, before the sixth session, seemed to have been given to the question of how the Enterprise would fare under a system of production limitation. If, up to a given limit, sites were allocated on a first-come-first-served basis, the Enterprise might not get any, if it were not quick off the mark. Conversely, if, within an agreed limit, it was insisted that the Enterprise should be assigned one site for every one exploited by a private or state contractor, that might heavily, and unacceptably, reduce total output from the Area, even in an extreme case (where the Enterprise failed to exploit any of its assigned sites) to the point of halving it. The solution adopted, which was embodied in the second and subsequent revisions of the ICNT, was to reserve to the Enterprise an amount of 38,000 tons of nickel a year (the expected yield of a single site). (46)

The Enterprise was thus to be furnished directly and through the obligations of contractors with substantial advantages in order to enable it to compete with the private and state sector. It could be seen as an "infant industry" to be protected initially until it achieves the momentum necessary to survive and prosper. It could also be seen as an investment in global political cohesion. It is more difficult to make a global organisation into a successful entrepreneur, than to limit it to the role of manager and tax-collector for mankind vis-a-vis the enterprises of others. By 1980 there was general agreement that the attempt should be made, and that, in particular, the money would be found, running to $1 billion in total liabilities, to get the Enterprise launched, in return for "access", that is a clear-cut indication in the convention as to the conditions under which states and companies could rely on being authorised to mine.

The parallel system was an ingenious attempt to find a compromise capable of operating equitably in an uncertain and largely unforeseeable milieu. In the absence of general confidence in the discretion of a sea-bed authority, it was bound to have its rigidities. It could not easily adjust the balance between the private (and state) sector and the international sector in the light of the new knowledge that becomes available from time to time, including evidence about the efficiency, impartiality and integrity of the sea-bed authority itself. Prudence dictated that arrangements be made for the operation of the system to be reviewed at suitable intervals.

There was a widespread understanding therefore, from 1977 onwards, that the duration of the "mixed system" (ie. the parallel system) would depend on the decision of a review conference, which would be convened automatically 20 or 25 years after the convention's entry into force. This was taken sufficiently seriously for it to form an essential element in the negotiations conducted by Jens Evensen in the Chairman's Working Group of the sixth session.

Proposals that the system of exploitation be consciously reviewed, by the Authority charged with administering it, go back to before Caracas (Conahan 1973), and at least one was made in the First Committee there. (47) The SNT would have provided for "the question of a general review of the provisions" of the convention,(48) but the United States, in December 1975, had called for the deletion of this provision, and this was one of the US demands which Engo, in the RSNT, accepted. Kissinger, though, made some mention of it in his intervention at the beginning of September 1976 and over the next year it absorbed, both at the sixth session in 1977 and at the preceding intersessionals, a great deal of attention.

A review could be conceived in two ways. Either it could be seen as an exploration of whether there were any changes which all concerned would see as improvements, in which case no-one would object to it; or it could be seen as capable of leading to changes that might be imposed by some on others, that would not require, before coming into effect, that every beneficiary of the existing system had given its consent. So conceived, the idea of a review became fraught with political significance. The question then arose, were the "untrammelled rights" sought for private and state applicants liable, in such a review, to be annulled?

By the end of the barren fifth session, in 1976, some were clearly seeing it in that light, indeed as heralding the end of the parallel system. Engo himself, in his final report to plenary on the work of his committee during that session, put it succinctly:

> "We have now come to confront the central and most difficult problem of all and it is this: should the new system of exploitation provide for guaranteed permanent role in sea-bed mineral exploitation for States, parties and private firms? Or should such a role for States, parties and private firms be considered only at the option of, and subject to conditions negotiated by, the Authority? Or again, should their role be conceived of as essentially temporary, to be phased out over a defined period agreed to beforehand?" (UNCLOS III, Official Records, Vol. VI 1976, p.132)

The proposal for a review thus became, not merely a way of enabling members to stand back periodically and monitor the way the system was going, and decide, in respect of those aspects of it which they already had the right to decide, what changes, if any, were necessary to set it on course towards generally accepted objectives; but also an attempt to shift, to the future, certain fundamental decisions on which agreement could not be reached now, on how long the dual system should last.

At the intersessional meetings at Geneva, under the chairmanship of Evensen, Mexico proposed two new articles (64 and 65). The second provided for a review conference, to be convened twenty years after the treaty provisionally or definitively came into force, which was committed to maintain the principle of the common heritage of mankind through an international regime and an Authority, but would be able to decide to amend the relevant part of the convention and, in so deciding, determine its own system of voting.

On May 13th 1977 Evensen circulated to the sixth session a somewhat amended version of these two articles as his own "informal suggestion for compromise formulation". They inserted passages into Article 65 which would have added to the principles that were to be sacrosanct, specified that the majorities for adopting such amendments should be not less than those required for decisions at UNCLOS III, with the same necessity to determine that "every effort to reach agreement ... by way of consensus" had been exhausted, and made clear that any changes resulting from the review conference would not affect existing contracts.

The industrialised countries expressed strong opposition to the idea that the review conference should provide an instrument for changing the system against their individual or collective wishes. The Soviet Union proposed to amend the articles drastically, and in particular to make it clear that any amendments it adopted could come into force only through the normal process by which amendments came into force. West Germany suggested that amendments should require ratification by three-quarters of the membership and by members representing three-quarters of the total production, and three-quarters of the total consumption of the categories of minerals derived from the area, (and that any member objecting to any such amendment be permitted to withdraw from the organisation). The United States emphasised the "impossibility" of the Senate ever agreeing to belong to an organisation whose character could be amended without the US's own consent, (despite American membership of the World Health Organisation which permits amendments by a two-thirds majority of the World Health Assembly).

Nevertheless, Evensen's third revision, circulated on June 11th, which was part of the package on which the United States, at any rate, was prepared to negotiate, did not, on the face of it, offer the industrialised countries the assurances they sought; it provided that the review conference should determine the system of voting and entry into force procedures, but that the majorities required under its voting system should be the same as those required for UNCLOS III, and while the reference to "entry into force procedures" implied that decisions of the conference would not, ipso facto, have the force of amendments, there was nothing to prevent the review conference from deciding that amendments should come into force when ratified by majorities no greater than the conference decisions required to adopt them.

On the other hand, the Evensen proposal implied that if no amendments came into force as a result of the review conference, the existing system should continue, and developing countries, expecting that the "coming into force" procedure would be made so difficult that the conference could not become a vehicle for radical change, wanted to include some element of incentive for such change in the treaty. Engo responded to these demands by a drastic alteration of the Evensen text, in which, if the review conference failed to reach agreement on whether or not those parts of the treaty relating to the sea-bed mining regime were to be amended, the private sector only was to be replaced by a system of joint ventures. (49)

The change this represented from the Evensen text was one of the major sources of American dissatisfaction with the ICNT. Much of the time, in the next round of intersessional negotiations, was taken up with it, and the seventh session in 1978, when it eventually got down to substantive business, included the whole question of a review conference in the mandate of the first of its seven negotiating groups, under the chairmanship of Frank Njenga of Kenya.

By the end of the Geneva part of the seventh session there seemed to be general agreement that a solution, if not already found, was considerably nearer than it was before. Njenga's own proposal took up a suggestion which had been made by Brazil at the previous session that in the event of no agreement at the review conference after five years, the Assembly could decide, without prejudice to existing contracts, that no new contracts, or plans of work by the Enterprise, should be approved. In this way he hoped to put pressure on all parties to work towards agreement of some kind, to continue the system either unchanged or with such amendments as were adopted.

Richardson, for the United States, recognised that this constituted a significant improvement on the ICNT, but warned that his country "could not agree to the possible termination of its right to access to deep sea-bed minerals at the time the need for them may become more acute". Over the next two years, Richardson, and his colleagues from other mining states, came to accept what they had declared unacceptable in 1977, the possibility of the review conference leading to the adoption of amendments by a two-thirds majority which would come into force when ratified by a two-thirds majority of states parties. A provision to this effect was inserted into the second revision of the ICNT in 1980. (50) In return, the developed states had been given the assurance that "the rights of states" would be among the features of the system that would be preserved intact by the review. Exactly what this meant would have given room for much debate. It might have meant anything; that the review conference majority would simply have made whatever changes it wished, and ratified them, regardless of the assurance. It was not likely to mean that

169

the parallel system as embodied in the convention was inviolate. It might, however, have ensured that some kind of parallel system survived, though perhaps a rather different one from the one just described; and it was made clear that any such amendments would not affect existing contracts.

THE REAGAN REVIEW

As we have seen, the quasi-compromise of August 1980 won widespread acceptance at the time, and such complaints as there were mostly concerned not the system of exploitation as defined in this chapter, but the effect of the production limitation (several European mining states saw it as too restrictive; the Africans, mostly, as not restrictive enough) and the adequacy of representation of the smaller developed states on the Council. Nevertheless it was clearly incomplete. The participation of international organisations (like the EEC) and of liberation movements (like the PLO), the structure, power and decision-making system of the Preparatory Commission which, after signature, would prepare the ground so that everything was ready for its coming into force, and the necessity of giving some recognition - "grandfather rights" - to investors who might be given title, in the meantime, by the legislation that had just been passed in the USA and West Germany, all required further negotiations, leading to texts to be somehow incorporated into or adopted with the draft convention before it could be offered for signature. These "leftovers" will be discussed in more detail in a later chapter. They were dwarfed in the event by the abrupt actions of the Reagan Administration on the eve of the tenth session in sacking several members of the Richardson team and reinstating Leigh Ratiner, and announcing a comprehensive review of America's law of the sea policy which would not be completed in time for the resumption of the session at Geneva in August. Ratiner had spent the Carter years advising Kennecott, and took a much more critical view of the emerging system of exploitation than Richardson. In 1978, for instance, he described the texts produced at Geneva in the spring of last year, which Richardson had seen as "the first ... to come to grips with many of the seemingly intractable issues associated with deep sea-bed mining" (Richardson 1978, p.4) as "showing a few minor improvements [over] and, in some important respects, worse than" what Richardson himself seen as the "fundamentally unacceptable" ICNT. (51)

When President Reagan announced on 29 January 1982 that the USA would return to the negotiations, the likelihood was, then, that it would seek major changes in the 1980 compromise, now before the conference in another version (the 1981 Draft Convention). A fuller idea of the specific concerns of the US government was revealed on February 24, in a document listing twenty problems. Of these twelve concerned the system of exploitation covered by this chapter, three decision-making in the Authority, three production limitation, and two the not yet negotiated issues of participation and preliminary investment protection. Of the twelve, four concerned access, two the review conference, two the question of technology transfer, and four the status of the Enterprise. However, in many of the twelve areas classified here as pertaining to this chapter, the problems were seen as concerned with the structure of the Authority (including the Enterprise).

The document claimed that because "the real power to grant or deny access rests in the Legal and Technical Commission", its composition should be limited to "mining and related technical experts" and not subject to the requirement of "equitable geographic distribution", (or that some other, 'automatic', system of granting contracts replace the present one); it was also concerned about the Authority's capacity to make and interpret rules. It distrusted an international inspectorate; it thought that the Council's unanimity rule would deter mining both because it would result in broad regulations which permitted the Authority too much discretion, and because outdated rules would not easily be repealed. It objected to an incomplete list of the fields in which the Council could make rules, and to the Authority's jurisdiction extending beyond mining to cover marketing and commodity agreements. Apart from the concern about the composition of the Legal and Technical Commission (which will be looked at in chapter 9 below), these worries seem exaggerated, and even contradictory, given the unanimity rule on rule-making, and the contractor's access to dispute-settlement procedures. Concern was also expressed that the convention as it stood would deter the development of sea-bed resources other than nodules, which by its terms would have to wait until appropriate rules, regulations and procedures had been devised. The US wanted the convention to recognise the freedom to explore and exploit these resources pending the adoption of a suitable international system of exploitation.

There was also strong opposition to the provisions for the mandatory transfer of technology; the simplest remedy for this was seen as the replacement of the provisions by "a combination of joint and individual state responsibilities to assist the Enterprise to obtain technology" (for which as we have seen, the convention to some extent already provided); but the document did not rule out the survival of the technology-transfer obligation of contractors, provided that 1) the Enterprise had to convince a "neutral" body that it could not acquire the technology elsewhere, 2) it applied only to the contractor's own technology, not to that of a third party, 3) it could not be transferred by the Enterprise to developing countries, and 4) "objective standards" were set for "terms and conditions of transfer", and 5) contractors could be forbidden by their government to transfer technology which was important for security, and would not be penalised by the Authority for complying with such prohibitions, thus indicating that the Administration was prepared to place more trust, on security matters, in firms, than in an international organisation like the Sea-bed Authority.

On the review conference, the US was not surprisingly worried by the possibilities of amendments being adopted, and coming into effect, with the support of any two-thirds of the parties; they sought to make the requirements for adoption, or ratification, of amendments more onerous, so as to give a vital role to states who had sponsored existing contracts and major producers and consumers. The proviso that such amendments should not affect rights acquired under existing contracts, was also felt to be inadequate. (52)

With regard to the Enterprise, the idea was partly to change decision-making so as to reflect the views of its creditors (a matter for chapter 9), partly to set out a strict schedule for the repayment of the monies lent to it, and partly to redress what were seen as advantages enjoyed by it over the private sector. The US wanted the Enterprise to be subject to the same production policy, and to have only temporary exemption from financial payments; it sought to abolish its right to have personnel trained by contractors; and, fearful that the Enterprise might exercise a monopoly over second generation sites, it wanted it to be required to release its rights over reserved areas in the absence of plans to work them within a given period. It even suggested - with scant regard to the interests of states in which sea-bed mining was a state activity - that the Enterprise might be required to conduct its operations through joint ventures.

The only important areas untouched by these criticisms were that concerning the financial obligations of contractors and the "anti-monopoly" clauses.

These "approaches" were put before delegates in February before the US delegation had received its instructions (Ratiner 1982, p.1009). When these had been received, they were translated into specific proposals for amendment issued as a Green Book early in March. The major changes in the system of access would have been (1) the streamlining of applications so that their disapproval required consensus in a specialist Technical Subcommission (Annex III, Article 6.2 bis); (2) the giving of ultimate rule-making power to a majority of the seven states assessed for the largest UN contributors (Article 161.7 c ter), (c) the restriction of the definition of the resources to which Part XI should apply to those for which Rules, Regulations and Procedures had been adopted, so that until that happened the exploitation of minerals other than nodules would not fall within the Authority's scope (Article 133) and (d) the abolition of the technology transfer requirement (Annex III, Article 5), with the substitution of a clause committing contracting or sponsoring states to ensure that the Enterprise can buy the technology it needs (Annex IV, Article 11 bis).

The chief change in the position of the Enterprise was a proposal that the "reserved" sites it acquires would revert to being generally available if not used within a few years, either for joint ventures, with the contractors, or some other qualified entity, or for simple exploitation by the original contractor.

The permanence of the resulting system was to be reinforced by providing that any amendments from the review conference, before coming into force, had to be ratified by all members of the Council as well as two-thirds of all members. (a further amendment - belonging to chapter 9 of this book - would have ensured that the Council always included the seven states parties with the greatest contribution to the UN budget. It would thus have been possible, under this proposal, for amendments to still have been carried over the heads of an ocean mining state like Belgium if it did not happen to enjoy Council membership at the time).

A group of eleven lesser developed states, most of whom either had no immediate prospect of engaging in ocean mining, or were already exporters of nodule minerals, calling themselves "the Friends of the Conference" and sometimes referred to as the "good samaritans", proposed a series of lesser changes.

These proposals would have left a consent element in the technology transfer mechanism: a contractor would have had to transfer to the Enterprise only the technology he was prepared to transfer to other third parties and on the same terms. They would have allowed the Legal and Technical Commission to disapprove applications by a 75% majority. The Authority was to be empowered to make rules applicable to non-nodule resources, but Article 133 was left unchanged so that without such rules such resources would be ineligible for exploitation. A curious provision was suggested for the review conference whereby if a state party did not accept an amendment adopted and ratified by the prescribed two-thirds majority, although it would come into effect, that member would not be bound by it and could still enjoy the rights and be subject to the duties of the unamended convention. The Group of 77 was willing to consider these changes, as a basis for negotiation, only if they were exhaustive: the USA attached such strong reservations to its response to them as to rule out much hope of reaching agreement by that means (Ratiner 1982, p.1016).

Thus although the Group of Eleven proposals were commended to the First Committee by its chairman, Engo, negotiations were never officially entered into on that, or any other basis, between the United States, and the Group of 77. The USA did, however succeed in winning the support of the other ocean mining states for a series of amendments to part XI of the draft convention, (53) which fell somewhat short of the original demands of the Green Book. Among the plethora of changes which that document had called for, the two crucial areas, according to Leigh Ratiner, were those of transfer of technology and the review conference. In respect of the review conference, the seven-power proposals were even more extreme than those of the US Green Book. Amendments adopted by such a conference would require, in order to come into force, ratification by all States Parties, not just two thirds of the States Parties including all members of the Council; thus even the Belgians would enjoy a veto over amendments. The seven powers did not, however, propose to amend Article 315, whereby amendments can be proposed at any time and are considered adopted or approved by the Council and the Assembly. Thus the proposal would have put more obstacles in the way of amendment after a review conference had run its course, than the convention otherwise provided for amendments that might be made at any time.

On the transfer of technology, the seven-power amendments were less drastic than the US Green Book. The latter had proposed to eliminate the entire article, in the Basic Conditions of Exploitation Annex, in which the contractor's obligation in this respect had been set out. In L121, the obligation was to remain, but be confined to technology which the contractor was able and willing to make available to third parties; this the contractor was to offer the Enterprise on no less favourable terms. The contractor was also required to advise the Enterprise, on its request, on the purchase of technology on the open market. In this article the seven mining states closely followed the proposal of the Group of 11.

The amendments of the seven, which in all would have affected 32 articles, also embodied the US proposal for limiting the period for which the Enterprise could hold "reserved" sites without exploiting them, a proposal widely seen as pernicious among the champions of the Enterprise because they struck at the root of the possibility that that body would steadily accumulate power and resources as the system blossomed. Of the other changes proposed by the seven most either fell within the scope of other chapters of this book, or were vague in what they entailed, such as one making the award of contracts more automatic, and one giving states a bigger role in ensuring the compliance of contractors with the Authority's rules. One exception which, though not incorporated in the convention as adopted, seemed to attract little attention, would have imposed heavier rather than lighter burdens on contractors. By what would have been a second paragraph of Article 138, signatories to the treaty would have agreed to enforce in the Area the standards of working conditions and maritime safety laid down by the ILO and IMCO.

On 23 April, President Koh, with the support of the Group of 77, was able to propose three amendments to Part XI and its associated annexes, followed by a further five on April 29. The most concrete of these changes were concerned with the review conference. It was to follow the same decision-making procedures as UNCLOS III itself; and any amendments it ultimately adopted and also any other amendment to Part XI of the convention (but not to Annex III or IV) would require ratification by three-quarters rather than two-thirds of all states parties in order to come into force. For the rest, the Koh proposals were largely verbal. "The development of the resources of the Area" became the first objective of the Authority's policy, in Article 150; and the procedure for handling applications for contracts was streamlined. The technology transfer article remained untouched.

In the end, the seven-power set of amendments, like all but those of the other amendments formally proposed, were never put to a vote. The gulf between the potential ocean mining states and the Group of 77 appeared wide. American opposition, and abstentions by other important states, primarily on the grounds of the system of exploitation, marred the adoption of the convention on April 30 1982. Nevertheless Leigh Ratiner believed that agreement was close

"if the United States had not demanded virtually autocratic ruling powers over the Sea-bed Authority and had not sought the total elimination of the so-called production ceiling ... it would have been possible, perhaps easy, to have obtained major improvements in the practical effects of, and the principles contained in, the technology transfer provisions and the provisions which permit amendments to the treaty (after 20 years) without US consent. When the US position did reflect modest relaxation in some of these areas, the negotiations came very close to solutions for these latter two problems" (Ratiner 1982, p.1014).

He was not alone in that belief.

NOTES

1. This could lead to "nodule wars" like the "railway wars" of the nineteenth and early twentieth century, where a second railway company, by buying land it never intended to develop, could extract a large sum from the original company in return for promising not to use it to build a rival line.

2. See chapter 5 above.

3. General Assembly Resolution 2573D, of December 15, 1969.

4. General Assembly Resolution 2749 of December 17, 1970.

5. A/AC 138/25 of August 3 1970 (USA); A/AC 138/26 of August 3 1970 and A/AC 138/46 of August 3 1971 (the UK); A/AC 138/27 of August 4 1970 (France); A/AC 138/33 of July 20 1971 (Tanzania); A/AC 138/43 of July 22 1971 (USSR); A/AC 138/44 of August 3 1971 (Poland); A/AC 138/49 of August 10 1971 (Chile, Columbia, Ecuador, El Salvador, Guatemala, Guyana, Jamaica, Mexico, Panama, Peru, Trinidad and Tobago, Uruguay and Venezuela); A/AC 138/53 of August 5 1971 (Malta); A/AC 138/55 of August 20 1971 (Afghanistan, Austria, Belgium, Hungary, Nepal, Netherlands, and Singapore); A/AC 138/59 of August 24 1971 (Canada); and A/AC 138/63 of November 23 1971 (Japan).

6. Developing countries made up 60 of the 62 votes in favour; developed 27 of the 28 votes against. See Buzan 1976, p.100.

7. "Since ocean space is a single ecological system and since its uses are increasingly interlinked, the new international order must reflect a total and comprehensive approach to the marine environment and to the international regulation of its problems" (A/AC 138/53, p.5).

8. Including Leigh Ratiner, the chief US delegate to the First Committee at Caracas, who was to join with the USSR in ruling out the idea but had previously, in his personal capacity, endorsed it at the 1973 conference of the Law of the Sea Institute (Gamble and Pontecorvo 1973, p.11).

9. A/AC 138/27, Part IV.

10. See A/AC 138/43, for the USSR proposal, notably Article 8 and the then blank Articles 9 and 14; A/AC 138/44, Articles 1 & 4, for the Polish one.

11. A/AC 138/49, Article 16.

12. A/AC 138/33, Article 34(2). This would have been a commendably egalitarian proposal, always assuming that it was defined in terms of benefits per head related to UN contributions per head.

13. Discussed in the previous chapter.

14. A/AC 138/25, Annex A.

15. ibid., Annex D.

16. A/AC 138/63, Article 36(c).

17. A/AC 138/59.

18. A/AC 138/43, Article 12.

19. ibid., Article 10, and A/AC 138/44 II, 2(b).

20. A/AC 138/63, Article 20.

21. A/AC 138/26, Article 8 and A/AC 138/46, Article 7. This scheme was ridiculed as liable to lead to a patchwork quilt of blocks governed by different states under national rules. It would though, have had the merit of conserving some of the best blocks for future generations.

22. A/AC 138/27, Part I.

23. The Polish proposal also allowed for the possibility that the organisation's role might develop, so that, at a later stage, it might graduate to 'supervision' and 'regulation' of sea-bed activities.

24. Uncharacteristically, the Soviet proposal did conceive of the possibility of expulsion of members in certain circumstances.

25. Though heralded by Christopher Pinto as 'a matter of the highest importance' when first formulated in the working group at Caracas, this did not perceptibly differ in substance from the 1971 Latin American/Caribbean proposal.

26. A/Conf. 62/C.1/L7.

27. For Belgium and the Netherlands, who had joined in sponsoring the proposal of Afghanistan and other land-locked and geographically disadvantaged states in 1971, this meant a change to a less flexible and more homogeneous system.

28. Apart from a requirement in the EEC proposal that no applicant should hold more than six contracts at any one time in respect of each category of resources.

29. Though Kissinger announced the United States' acceptance of the dual system as such, that is of the Authority's right to engage in exploitation itself so long as companies (and states) were also assured of that right, the previous August (Kissinger 1975).

30. A/Conf. 62/C.1/SR20 p.5).

31. Leigh Ratiner, in an "extemporaneous" speech to the working group of First Committee of August 9, 1974, gave a most trenchant exposition of this view.

32. Though the US amendments to the SNT, proposed in December 1975, included a proposal to delete the list of categories of mineral resource.

33. The analysis that follows takes these questions in

reverse order, which might be said to be the ascending order of difficulty of resolution.

34. See, for instance, Smith and Wells 1975, and Smith 1976. Britain's revision, in August 1978, of the taxation arrangements for companies exploiting its North Sea oil, is a piquant case in point.

35. SNT (A/Conf. 62/W.P.8) Part I, Annex I, para. 15.

36. A/Conf. 62/122, Article 162 2(w).

37. ibid., Annex III, Article 18.1.

38. A/Conf. 62/Cl/WR.5/Add 1, para. 2.

39. See chapter 2 above.

40. Thus an applicant envisaging an operation that would yield three million tons of nodules a year for twenty years would have had to pay $ 60 million within a month of the signing of his contract.

41. "A Cost Model of Deep Ocean Mining and Associated Regulatory Issues" cited in "The Chairman's Explanatory Memorandum on Document NG2/7", in UNCLOS III, Official Records, Vol. X 1979, pp. 63-69. (This appears to be the model described in Miles and Gamble 1977.)

42. A./CONF. 62/25 Table II, p.76.

43. They included a version of the Indian "charge to mine" ($1m a year); and attributed 40% of the value of metals to the mining stage. Their net effect would be that those paying by production charge only would pay 7.5% in the first 6 years; 10% in years 7-12, and 14% in years 13-20. In a mixed system, the production charge in these periods would be 2%, 4% and 6% respectively, and the profits tax 16%, 28% and 32% of total proceeds (40%, 70% and 80% of attributed net proceeds') with the proviso that the rate could not rise from 16% to 28% until the contractor had fully recouped his development costs, or from 28% to 32% until he had recouped them twice over.

44. Law of the Sea Convention, Annex III Article 8.

45. United States Mission to the United Nations: Press Release USUN - 97(7) September 1, 1976.

46. ICNT, Rev. 2 Article 151 2(c).

47. See my own intervention in the committee's sixth meeting, UNCLOS III, Records, Vol. II 1975, pp.25-26.

48. i.e. of the part concerned with sea-bed mining; it was then thought possible that a separate convention might be

adopted on the First Committee's mandate in advance of a comprehensive law of the sea convention.

49. See chapter 5 above.

50. This recourse to a two-thirds majority could be invoked only in the sixth year of a review conference, convened fifteen years after the earliest commercial production had begun in the Area, and only if the preceding five years of deliberations had failed to result in agreement (ICNT Rev 2, Article 155.5).

51. Letter to Congressmen, May 26 1978, attaching copy of Dubs 1978.

52. It is not clear why, and in the subsequent Green Book of specific amendments no change in it is proposed.

53. A/Conf. 62/L121 of April 13 1982.

8 Production Control

Throughout the process of negotiating a sea-bed authority, production control, as an issue, has tended to be conceived of as a means, and the most effective one, of protecting land-based producers of sea-bed minerals from excessively drastic impact of sea-bed competition; but the protection of such producers is not the only reason for giving an international sea-bed authority the right to control production; nor is the control of production the only way of protecting them.

Given how rudimentary our present knowledge of the environmental consequences of deep sea-bed mining is,(1) there are environmental reasons for controlling the rate of production; the slower it proceeds, the less damage will have been done before any environmental (or other) hazard is detected and appropriate action taken to guard against it. The delicate balance between adventurousness and prudence should not be left to mining enterprises alone, who would have no incentive to avoid environmental damage.

Moreover if the Authority were to enjoy the discretion to fix, or negotiate, the financial terms of contracts, scarcity would improve its bargaining power.

If there were more bidders than permitted sites, it could exact, for globally-approved purposes, something approaching the maximum "take" for each site consistent with the economic profitability of mining it.

The main losers from sea-bed production would be the countries that are, or might be, land-based producers, and net exporters,(2) of the same or similar minerals as those that might be obtained from the sea-bed. In the case of manganese nodules these would include existing exporters of copper, nickel, cobalt, and manganese; potential exporters, i.e those which had deposits of any of these which might in the absence of sea-bed mining have been exploited for export in the foreseeable future, and actual and potential exporters of products competitive with any of these, such as aluminium. Land-based companies would also be affected, particularly those that were heavily committed to existing sites. Those that could move would increase their bargaining power vis-a-vis site owners, but worsen it vis-a-vis their customers.

Clearly, to the extent that it occurs, sea-bed production will make the price of the refined metals in question lower than it would otherwise be. That is, it may dampen or eliminate an absolute rise, transform a rise or a stable price into a fall, or accentuate an existing fall. It may be that all consumers of products using copper or nickel, and thus virtually all countries, will benefit from the resulting increased supplies, and lower prices, of these metals, on balance, though, as between net exporters of primary products and raw materials and net exporters of manufactured goods, a fall in the price of the former will alter the terms of trade in favour of the latter, and thus against the former.(3)

Moreover, the closure of land-based mines can leave communities devastated, and resources - labour, social capital and infrastructure - unused. Such communities need time to adjust.

Given, then, a fairly widespread recognition at UNCLOS III that the level of production from the international area of the sea-bed should not be left to be determined solely by "the market", the proposed remedies have tended to take two forms: production limitation, and compensation. Compensation schemes are not easy to administer. There is much scope for debate over the question of how much damage land-based producers have suffered through sea-bed mining; and it is difficult to collect compensation for the real beneficiary of change, the consumer; but such schemes do have the advantage, compared with production limitation, that their benefits can be confined to the deserving; and they can be used to cushion change, rather than simply arrest it.

With the possible exception of the environmentalist argument, where United States' interests are ambivalent,(4) all these arguments for production control, or for the compensation of the victims of sea-bed competition, run directly counter to the interests of the United States. She is net importer of sea-bed minerals rather than a net exporter; a pioneer rather than a late-comer to sea-bed mining; and a likely contributor to, rather than a beneficiary of, any redistribution of wealth that the Authority might ever effect. For the US to support, or even tolerate, any form of production control, would have required either altruism, or far-sightedness, or a keenness for a treaty for the sale of the other interests that it would serve. In fact, from the start, it has been one of the staunchest opponents of such control.

Furthermore, as we have seen, the US, and other mineral-importing developed states, were apprehensive that any power the Authority had to control sea-bed mining might be used not merely to regulate production, but to make it quite unprofitable. They did not trust the body that seemed likely to emerge, and were therefore very reluctant to give it any discretion. Thus when in the end the US did concede the principle of a limit on production, it insisted that the means by which the limit was to be calculated should be written into the treaty.

Production control, as an issue, was not originally distinguished from the rest of the debate over whether a sea-bed authority should be established and, if so, what its functions and powers should be. Dr. Pardo, in his original address to the General Assembly's First Committee in 1967, did not specifically refer to the damage sea-bed mining might do to the economies of land-based producers, still less to measures by which damage might be averted or minimised; but he did call for the creation of an agency "endowed with wide powers to regulate supervise and control all activities on or under the oceans and the ocean floor", and saw the regulation of commercial exploitation as one of these powers (Pardo 1975, p.39).

Concern for the effects of sea-bed mining on land-based producers was first raised by Argentina, in the Economic and Technical Working Group of the Ad Hoc Sea-bed Committee, in 1968 (Oda 1977, p.20). Argentina's point was taken up by that group in its programme of work and won support from at least one importing country, Japan (Oda 1977, pp.31-32). In the General Assembly's Declaration of Principles of 1970, there was mention, albeit only in the preamble, of the duty to "minimize the adverse economic effects caused by fluctuation of prices of raw materials resulting from such activities", and Resolution 2750 A, adopted at the same time, called on the Secretariat to study such problems, and produce two reports, one in conjunction with UNCTAD and other specialised agencies. The Secretary-General's preliminary report, in 1971, 67 pages long in 1971, was optimistic, except in respect of manganese producers. A further report the following year added cobalt producers to the danger list and saw a clear need for control by the Authority. These views were reiterated when in May 1974, the Secretariat's main response to Resolution 2750A was published.

The report was not alarmist. It expected only a minimal impact of sea-bed mining on the copper market; and for nickel, assuming, optimistically as it proved, a continued 6% annual long-term growth in demand, it did not anticipate serious damage to land-based producers. It did stress, though, that the regime established should have flexibility and suggested that "metal production from nodules be geared to supplying between 50 and 100 per cent of the increase in demand for nickel". Finally, it saw the threat to the cobalt and manganese markets, while substantial, as capable of being met, as far as developing country exporters were concerned, by 'the compensatory approach'.

Three of the eleven proposals of 1970-71, by states and groups of states, contained explicit provision for production control: those of Canada, Chile and others, and Tanzania. There was also some general reference to it in the Polish, Japanese, and Maltese proposals and that of Afghanistan and other land-locked and geographically-disadvantaged states. Others envisaged creating organs of the Authority specifically for this purpose: a "Resource Management Commission" (Canada), a "Stabilization Board" (Tanzania), and a "Planning Commission" (Chile and others).

When UNCLOS III met at Caracas, the perceived need for such proposals had been enhanced not only by the Secretary-General's own reports, but by several studies produced by UNCTAD, which saw developing land-based producers as likely to lose export earnings (compared with what they might have earned) to the extent of $360m a year or more by 1980, well beyond what the Authority would be in a position to compensate. As a result ocean mining "while contributing to world economic development, may also result in a widening of the income gap between developed and developing countries".(5)

Moreover, the Caracas session took place in the immediate aftermath of the fivefold increase in oil prices and the starkly confrontational sixth special session of the General Assembly which adopted the Declaration and Action Programme on a New International Economic Order. These events enormously encouraged mineral exporters; but also stiffened the resolve of potential ocean miners, and mineral consumers, not to let themselves be denied access to what they saw as a means of keeping some primary product prices down to "reasonable" levels. Following addresses by representatives of UNCTAD (Mr. Arsenis) and the Secretary-General (Mr. Stavropoulos and Mr. Branco) on this subject, and question-and-answer sessions (to which the US complained that "delegates should have the opportunity to hear other answers"), the First Committee devoted two evenings, early in August 1974, to "informal seminars", addressed supposedly by "experts" rather than "delegates", e.g. to the question of the "economic implications" of sea-bed mining. There were signs that this pressure was affecting American policy. Whereas an earlier US Working Paper of July 31 had virtually dismissed the problem, by August 8th Leigh Ratiner was prepared to admit that "A solution to the problem of developing country producers must be found" though such a solution had to maintain "a balance among all economic interests", and suggested "an accommodation which provides a mechanism for reviewing, on a continual basis, whether the problem does indeed exist, what are its true dimensions and what measures could most appropriately be enacted for its resolution in the light of conditions existing at the time."(6)

No such mechanism, however, was visible in the US proposals for "basic conditions of exploitation", produced at the same time,(7) which seemed firmly to preclude giving the Authority the discretion to impose any restrictions on the areas available for exploitation; an attitude shared by Japan (8) and by the eight EEC countries that gave their names to L8.(9)

By contrast, the Group of 77's version of these "basic conditions", (10) though it did not specifically refer to the economic consequences of sea-bed mining, would certainly have furnished the Authority with the powers necessary to control them. These included the powers to determine the part(s) of the Area to be opened up for exploration and exploitation (para. 3); to enter into contracts only when it considered it appropriate (para. 5); to retain title to the minerals through all stages of operations (para. 2); and, in spite of a provision purporting to ensure security of tenure (para.10), to revise, suspend or terminate a contract "in case of a radical change in circumstances".

The Third (Geneva) session made little progress, if any, in resolving this issue. Engo's Single Negotiating Text would simply have given the Authority, through an Economic Planning Commission, the power to determine what parts of the Area would be exploited and how fast. Pinto's unoffical text was similar, with the important difference that it would have so constituted the Economic Planning Commission that majorities of both exporting and importing states would have been required for its decisions. Under such a system, if private and state enterprises were likely to be quicker of the mark than the Authority itself, which was generally assumed, restrictions on production imposed to protect land-based producers would fall more heavily on the Authority than on private applicants. The dilemma this posed was delineated in the last chapter. (11)

This difficulty did not show itself at Geneva in 1975 since the United States, in particular, continued to reject any production limitation until well into 1976. The amendments it produced to the SNT, in December 1975, would have emasculated, as well as further paralysed, the Economic Planning Commission. (12) Quiet diplomacy, before the fourth session, paved the way towards something of a compromise, at least as far as the leading land-based producers were concerned. (13) In return for acceptance of the parallel system, the United States was prepared to see a production limit written into the treaty. What this meant was made public in a speech by Kissinger of April 8th 1976 (Kissinger 1976). This speech, besides conceding a temporary limitation on sea-bed production, "tied to the projected growth of the world nickel market, currently estimated to be about 6% a year", also spoke of using some of the Authority's revenues, and those of other bodies like the World Bank, to compensate countries hit by sea-bed production and to help them diversify their economies or improve their competitiveness; and there was even some acknowledgement that the Authority might be empowered to participate in international commodity agreements (but only in respect of the amount of production for which it was directly responsible).

This did not mean giving the Authority _discretion_ to control the rate of production, since the limit was to be specified in the treaty; moreover, it offered only temporary protection to nickel producers, and since (as it transpired), the figure of 6% was to be a minimum in calculating the limit, this protection would be eroded to the extent that actual growth in nickel demand fell below this. In any case, sea-bed production was not expected to reach the proposed limit. Despite all this, it represented a distinct break with previous official policy; and while the limit would offer little comfort to cobalt and manganese exporters, the developing countries among them might hope to gain from the compensation provisions, although full compensation to both sets of exporters might even then have cost something approaching $100 million, a large slice of the Authority's likely income.

The reception of the Kissinger proposals was by no means enthusiastic. Land-based producers, and expecially _potential_ producers, wanted a more stringent limitation; As the Peruvian delegate said at a seminar in 1978

"In Peru, for example, nickel has been discovered recently. Yet, unless sea-bed production is regulated by an international regime, this Peruvian nickel may never be exploited. This is a prospect which Peru does not relish" (Quaker Office at the United Nations 1977, p.7).

Such producers therefore wanted the limitation to apply to less than the whole of the growth segment, so as to leave room for the possibility of growth of production on land.

At the fourth session, though, they seemed ready to settle for a bargain whereby, in exchange for the protection these proposals offered them, they would accept a dual system; Accordingly, the provisions of the RSNT for the protection of land-based producers (Article 9 para 4) followed the US proposals fairly closely, listing the three techniques mentioned - commodity agreements, production limitation and compensation. On the first of these the RSNT went significantly further than Kissinger. It would have allowed the Authority to negotiate and participate in such agreements on behalf of _all_ production in the international area, not just that part of it that fell under its direct control. The production limit was to last for at most twenty-five years, and to be derived from the growth segment of the nickel market, provided that this should be at least 6%.(14) Compensation was to be left to the Authority's discretion (Article 28 para 2(xi)), via an Economic Planning Commission which it would appoint, which would be balanced in composition as between producers and consumers, and whose recommendations would require only an overall two-thirds majority.

The largely unproductive fifth session, at which the RSNT was repudiated by the Group of 77, ended with a decision to reserve the first two weeks of the next to First Committee matters, and a claim by Engo, that committee's chairman, that there was now a recognised common interest in both increased availability of raw materials from the sea-bed and in protecting land-based producers in one way or another (15). In the next nine months, the initiative in First Committee matters was largely taken by the Norwegian Jens Evensen, first in the inter-sessionals of February 28 to March 11 1977, and then, as chairman of the "chairman's working group" at the sixth session in New York. At that session the Group of 77 wanted the limit to be set at 50% of the (actual) growth segment. Evensen himself, in a proposal circulated on May 18th, suggested putting it at 100% of the growth segment for the first five years to which would be added 75% of the subsequent growth segment indefinitely thereafter; it was thus no longer to be temporary, but was to lapse only when superseded by a commodity agreement.

There was much discussion, aided by a Secretariat study, about how many sites different formulations of the limit would permit, which would depend, inter alia, on what the actual rate of growth of nickel turned out to be. Against this background, Evensen made further revisions to his formula for the production limit. The final version, issued on June 18th, provided that the growth segment would be calculated from January 1st 1980, rather than when commercial production began, which was expected (rightly!) to be much later, and the limit would be set at 100% of that segment for seven years rather than five, but thereafter at 66 2/3% rather than 75% of it.

Evensen also made verbal changes in two related aspects. The first left implicit, instead of explicit, the Authority's right to join in commodity agreements on behalf of all production in the Area. The second circumscribed more stringently the circumstances which would warrant compensation or other assistance to developing land-based producers.

In the ICNT Engo, besides changing Evensen's language somewhat, also amended his formula for the limit. It was still to start on January 1st 1980 and include all the growth segment of the next seven years; but the addition for subsequent years was to be 60% of the subsequent growth segment rather than 66 2/3%. This fairly modest change, because it did not rise out of the negotiations, was one of Richardson's major objections to Part XI of the ICNT. Potentially more importantly, the ICNT also provided that "The Authority may regulate production of minerals from the Area, other than minerals from nodules, under such conditions and applying such methods as may be appropriate" (Article 150, para.I(c).

This clause showed foresight. Delegates had, in drafting the regime for the Area, tended to forget that nodules were not necessarily the only valuable resource it would prove to contain, and the production limit was otherwise formulated entirely with reference to nodule mining. It was however, quite unacceptable to the USA, and a further cause for that country's dissatisfaction with the ICNT. For Richardson, Evensen's proposals in the field of production control "were by no means acceptable" though as "the product of an open discussion" involving all countries participating in UNCLOS III, he saw it as possibly "a basis for further negotiation breaking the impasse on sea-bed mining". He had earlier, in a closed meeting of the chairman's working group, used much stronger language against Evensen's revision of the RSNT's Article 9.

As we have seen, after a series of intersessional meetings, at which "no-one had attempted to defend the existing text as reflecting a true consensus", Richardson announced in January that he would "keep the door open" for further US participation in the conference. One reassuring fact, from the US point of view, was that there were to be more intersessional meetings in February at which the production ceiling would be one of the three major items for discussion. The seventh session reorganised negotiations, identifying seven "hard core" issues, each assigned to a "negotiating group". Three of these dealt with First Committee matters. That to do with production policy, Negotiating Group 1, was chaired by Frank Njenga of Kenya, and appeared to have achieved dramatic progress by the end of the Geneva portion of the session. Njenga's claim that his "revised suggested compromise formula" for Article 150 and other related articles "reflected a broad measure of agreement between the various interest-groups"(16) was largely confirmed, as far as the US was concerned, by its delegation report.

The apparent progress stemmed, essentially, from a tentative agreement between the United States and Canada. This had three advantages for those who might want to mine the Area. First, a time-limit was set to the "interim period"; it would start five years before commercial production began and, as in Kissinger's original proposal, last for 25 years, unless superseded by a commodity agreement before then. Secondly, following some highly technical work by a sub-group under the chairmanship of Alan Archer of the UK, the method of deriving the limit was made precise. The growth segment was to be calculated by extrapolating a geometrical average of annual percentage increases over the previous fifteen years, in world consumption of nickel; that is, in any given year, a trend line based on the average annual increase over the previous fifteen-year period would be used to project world consumption for every year in the "interim period". The growth segment at any point on this line would be the cumulative difference between the consumption figures thus derived for the _first_ year of that period and those for all years thereafter up to the point in question. Thirdly it was made clear that the Authority could regulate mining _other_ _than_ nodule mining _only_ by amendment of the convention. Given these gains, it is easier to see why the USA accepted that the _level_ of the ceiling should be even lower that in the ICNT, 100% for just the first five years of the interim period, and only 60% thereafter. On the face of it, though, this seemed a startling change.

Another explanatory factor was that, by this time, there was a glut in the relevant metals market, and thus a reduced need to look to the sea-bed for new sources of supply. Even without production control, fewer than seventeen sites were expected to be developed by 2000.(17) Land-based producers were becoming worried by the possibility that sea-bed mining would be subsidised by governments to avoid dependence on the specific suppliers, although the US Dept. of Defense saw no immediate need for this.(18)

Nevertheless, the tentative agreement was not upheld by Washington, and Richardson had to return to the next session and attempt to raise the ceiling to a larger fraction of the growth segment. He also had to deal with Soviet demands for an "anti-monopoly clause," which now received some support in the West. The lower the aggregate level of permitted production, the less willing America's allies were to countenance a scramble in which all the sites might go to American companies.

By April 1979, the US delegation was having to seek changes in the supposedly agreed formula, not only by increasing the percentage split (after the first five years of the interim period), but also by establishing a floor below which the annual permitted sea-bed production was not to fall, however slow the rate of growth of nickel consumption turned out to be, and by ensuring some flexibility in the production authorisation once given, i.e., a range within which variation on a year-to-year basis would be possible. The land-based producers by contrast sought to ensure that the markets for metals other than nickel could not be swamped by unrestricted recovery of nodules without extracting nickel therefrom (the "four metal" problem); and the Group of 77 also wanted the Enterprise to have reserved to it an adequate share of the permitted production.

From 1979 these questions were assigned to a sub-group of Negotiating Group 1, originally of 25, then further reduced to 10,(19) and chaired by Satya Nandan of Fiji.

The results did not appear until the second revision of the ICNT in April 1980. The 60-40 split had to remain, as a general percentage, but a "floor" was introduced, whereby, if the actual nickel growth rate fell below 3%, that figure was used to arrive at the production ceiling. It was, however, qualified by a "cap." The production limit was never to be so high as to permit, for a given year in the interim period, more than the actual projected growth segment to be taken from the sea-bed. Thus for the 25-year period, ignoring for the moment the possibility of commodity agreements, the limit was to be 100% of the actual growth segment for the first five years, plus an addition for the following twenty that would constitute 60% of the growth segment for those years where the rate of growth was 3% or above; if it was below that figure, then the limit was to be 60% of a notional segment based on a 3% trend line up to a maximum of 100% of the actual growth segment.

This represented some concession to the consumers and ocean miners, as did a further provision allowing year-to-year variations in actual production such as might exceed the authorised limit for a year by 8%, provided that such excesses did not occur three years running, or bring the aggregate amount of production above that permitted. Increases of from 8% to 20% would require a supplementary permit.

In return, the "four metal" problem was met. Contractors were not to be permitted to evade the production limit by processing additional nodules for metals other than nickel. This did not however meet the increasing fears of producers of cobalt, whose price had risen far above that of nickel. Sea-bed production had always threatened to eliminate this price gap, and it now seemed that that would mean that cobalt producers would lose far more than the Authority could hope to compensate them for. Nevertheless the new provision was a concession to the Group of 77, as was a provision (Article 151.5) reserving to the Enterprise 38,000 tons (presumably annually) of permitted production. Additionally a new system was introduced, apparently to general approval, for administering the production ceiling. Production authorisations were not to form part of contracts, but were to be applied for within five years of the expected start of commercial production. They were to be granted automatically so long as the limit was not exceeded; and if, in any given four-monthly 'round', the maintenance of the limit required a choice among such applicants, elaborate rules were laid down (in Annex III, para.7) by which the Authority was to be guided. Priority was to be given, inter alia, to applications from reserved sites.

Fears by some developed land-based producers, notably Australia and Canada, about the possible subsidisation of sea-bed production, began to be articulated even more strongly. It was even suggested that the entire production ceiling provision could have been abandoned if such subsidisation could have been banned. This proved unfeasible, and a more modest suggestion, applying the GATT rules and the GATT dispute settlement procedures to such charges, was eventually incorporated in the text of the convention in 1982.

The last two years of UNCLOS III, as we have seen, were mainly taken up with waiting for the results of the review initiated by President Reagan, and attempting to reconcile them with the expectations of the Group of 77 engendered by the 1980 text. The US ambassador's remark to the Geneva part of the tenth session, mentioning a "lingering impression that the convention would run counter to a policy of encouraging and promoting sea-bed resource development" confirmed that production control would be a major item on the Americans' list of proposed changes.

Nevertheless, the pressure was not all in one direction. Four African exporters of cobalt and manganese, Gabon, Zaire, Zambia and Zimbabwe, asked that the Secretary-General should make a further study of their plight. When such a study reiterated that their industries "could be seriously affected" by sea-bed production,(20) they then sought to give the Authority the power to keep such production lower in respect of these two metals than the limit based on nickel would imply.(21) These failed; but the Preparatory Commission was asked to study the problem and the possibility of establishing a compensation fund.

Meanwhile, as the US prepared to resume participation in UNCLOS III at its eleventh session in March 1982, its "Approaches" document of February 24 duly included "Production Limitations and Policies" as one of the seven major areas of the convention which were seen as unsatisfactory, and listed a plethora of objections to any form of limit, blandly claiming that "sea-bed mining will not compete with existing land-based nickel-mining investments" (p.29). It nevertheless confessed itself unable to suggest alternatives that would solve problems of both sea-bed and land-based miners. One suggested policy for the USA was to insist on deleting the limit entirely; another, to make it operable only for twenty years and then only when there had been a sharp and sustained fall in the aggregate price of the metals involved. The Green Book of specific proposals, when it came, supported total abolition of the limit, leaving only the prospect of adjustment assistance to protect land-based producers. In addition, the Authority's right to participate in commodity agreements was to have been restricted in its application to production by the Enterprise and then only when 'all interested parties', i.e. all ocean investors and major producers and consumers who were on the Council, had agreed.

Leigh Ratiner saw the demand for the total elimination of the production ceiling as fatal to the possibility of negotiating amendments to the convention that would meet the most serious US objection to it. It was, as he saw it, a 'cosmetiç' provision; it reassured land-based producers (at any rate those of nickel) that something would not happen which in his view (which, given his connections with ocean mining, was a knowledgeable one) would not happen anyway. Until, in mid-session, it was modified, it was impossible to negotiate changes in technology transfers and in the review clause. Ratiner speaks only of "moderate relaxation" in this respect, but claims this change was enough to bring the USA very close to agreement on these two latter points (Ratiner 1982, p.1014). According to the New York Times, it took a visit to Washington by the US Delegation leaders, at the beginning of April, to produce this change in US policy.

Once the USA had accepted the possibility that the convention would contain some production limit, it was able to agree with the other main ocean mining states (Belgium, France, Germany, Italy, Japan and the UK) on proposals for improving it, embodied in the extensive formal amendments submitted by these states on April 15 1982.(22) By these proposals, the allowed proportion of the growth segment of the nickel market would, for the first five years of commercial production, have risen by 2% a year. Thus it would have been 62% in the eleventh year of the interim period and 80% in the 20th and thereafter. The seven also agreed on limiting the Authority's power to enter into commodity agreements to production by the Enterprise; and on deleting the word "compensation" from Article 151, so that adversely-affected developing countries would be able to receive 'economic adjustment assistance' only. Finally, pioneer investors were to be assured of production authorisations even if this meant temporarily exceeding the limit.

The one change in the production policies of the convention that was accepted at the final session applied to mining other than nodules. Rules governing such mining were to be drawn up by the Council, in the same way as those concerned with the nodule mining, not by means of amending the convention. This reflected the paradoxical result that it was easier for a majority to override a minority by amending Part XI, which after the review conference, could, in the end, be achieved by a 3/4 majority at such a conference and the ratifications of three-quarters of the parties; than to make rules under the convention as it stood where every member of the Council had a veto.

In the event, although differences on production control resurfaced with the Reagan review and were never fully resolved, it was almost certainly not on this issue that the attempt to adopt a treaty by consensus floundered. The world economic situation made the actual production limit specified by the treaty seem irrelevant. Arguments persisted over whether the Authority should be able to participate in commodity agreements on behalf of all production in the area, but these too seemed academic. Perhaps the chief importance of the issue was its ideological appeal. In the end the West in general, and the USA in particular, seem to have neglected the chance to make concessions on it that would have cost nothing in return for concessions by the Group of 77 on questions that mattered more to the West. When that omission was repaired, there was too little time left to do a deal.

NOTES

1. See chapter 2 above.

2. Countries which could expect to remain net importers of such materials, even though they had or might develop exploitable deposits at home, would presumably gain more from cheaper imports than they would lose through the loss of income from the reduced price of their own mining products, though they could suffer some deflation or unemployment in mining regions.

3. The first argument assumes a non-zero sum game; the second, a zero-sum game. Economic activity is not, of course, a zero-sum game; but it has always been a game in which, while aggregate productivity per unit of cost (and thus aggregate wealth) can increase, individual economic units - firms, categories of workers, and states, can lose far more than they gain from the increase in available goods on the market.

4. US companies probably enjoy an advantage, compared with their rivals, in respect of their capacity to absorb the costs of avoiding environmental and other hazards; and there is a not negligible environmental lobby, and a substantial interest in preventing ocean mining from interfering with, or endangering, other uses of the sea; but it is by no means clear that these considerations would, even in themselves, have outweighed the costs of controlling the rate of exploitation except in areas where alternative US uses of the sea would be manifestly vulnerable to such mining.

5. Statement of Mr. Arsenis, special representative of UNCTAD's Secretary-General, before the First Committee, at its sixth meeting, July 16 1974, summarized in UNCLOS III, Records, Vol II 1975, pp.26-30.

6. UNCLOS III, Records, Vol. II 1975, p.65.

7. A/Conf. 62/C.1/L6.

8. A/Conf. 62/C.1/L9.

9. A/Conf. 62/C.1/L8.

10. A/Conf. 62/C.1/L7.

11. See chapter 7 above. The dilemma arises out of uncertainty, both as to the exact level of the limit, and as to how soon and on how large a scale the Enterprise will operate. Unless production is reserved to the latter, the limit may have been reached by the time it is ready to produce.

12. That is, would have given it few, if any, significant functions and, by its decision making system, minimized its chances of performing even those functions.

13. Such producers may also have been conscious of the dangers that the Metcalf Bill would pass Congress and bring with it the spectre of totally unlimited mining.

14. RSNT, Part 1, Annex I, para.21.

15. UNCLOS III, Records, Vol. VI 1976, p.133.

16. UNCLOS III, Records, Vol. X 1978, p.20.

17. US Congress 1978, Table VII.1, p.112.

18. Statement by David E. McGiffert, Assistant Secretary of Defense, to the House Committee on International Relations, July 25 1977, in US Congress 1978, p.88.

19. The ten states were, apparently, Canada, Chile, Cuba, Indonesia, Australia, US, UK, France, FRG, and Japan (US Delegation Report 1979, p.10).

20. A/Conf. 62/L84.

21. A/Conf. 62/L101.

22. A/Conf. 62/L121.

9 The Structure of the Authority

Considering their importance, questions about the structure of the Sea-bed Authority have occupied a surprisingly small fraction of UNCLOS III's time, especially at its first eight sessions. Such questions include what organs the Authority should have, how they should relate to each other, and the composition, jurisdiction and voting system of each. Thus, given a convention establishing a Sea-bed Authority and entrusting certain functions to it, "structure" determines what combinations of what categories of states can combine to exercise those functions, and gives each prospective member of it some indication of the chances that the Authority will do something to which that member strongly objects, or fail to do what that member would have wanted it to do. In other words "structure" tells a state whether the Authority is likely to be its ally or its opponent, its endorser or its critic - or neither. It does not happen, of course, that those that have the votes in international organisations can settle issues as they wish, except for purely internal ones, like elections to subsidiary organs or appointments of officials; but they can confer legitimacy or withhold it. "Structure" determines, at the very least, who can embarrass whom. For a Sea-bed Authority, however, more than that was at stake. So long as all the mining states came in, voting power could mean the power to regulate the mining of the international sea-bed and to dispose of the revenues accruing to it thereby. Voting power is not, however, always easy to assess. States may think that a given system will assure them of a majority and then find that, because of shifting alignments or changes in membership, they are in a minority. Those aware that this could happen will tend to play safe, and make decisions difficult for an organisation; they will be supported in this, of course, by those who see themselves condemned to a minority position with no foreseeable prospect of change. This chapter, then, is mainly concerned with the debate over the power of majorities: either weighted majorities as in the IMF and the World Bank, or numerical majorities of some categories of member states. How big do they have to be, and once they have been assembled, what can stand in their way?

Western, and particularly United States, views on these questions have changed beyond recognition since the early days of the United Nations. In negotiating the Charter itself, Britain and the United States had largely agreed with the Soviet Union that the world should be run by a small group of privileged states, principally themselves, and that the remainder should be given only those limited powers necessary to induce them to join the organisation and accept the hegemony of the few. They all supported the principle of creating a separate organ (the Security Council) in which a few major states would each enjoy a veto at least over enforcement, that is, the identification of "threats to the peace, breaches of the peace and acts of aggression" and the adoption of measures to deal with them. They differed over what else it should extend to, and Britain and the United States would have allowed more scope for majorities than would the Soviet Union in such matters as the mere discussion of questions as opposed to adopting resolutions; admitting, expelling, or suspending members; and appointing the Secretary-General; but there is no doubt that in Cordell Hull's words "our Government would not remain there a day without retaining its veto power". In the end, though, it was agreed, and embodied in the Charter, that all these kinds of decisions should require the concurrence of each of the five permanent members and that amendments to the Charter had to be ratified by each of them. (1)

In the next few years this consensus rapidly broke down. Supported by allies in Asia, Europe and Latin America, the United States became confident of its ability to win a majority in the United Nations on what seemed overwhelmingly the most important set of issues which came before it, those relating to the Cold War; as it did, it became increasingly impatient with all encumbrances the Charter had placed upon majority rule, even those with which it had agreed from the start and would have insisted on if necessary. The Soviet Union's use of the veto was attacked as improper, and provoked by repeated proposals known to be unacceptable to it; the status of the Security Council was denigrated; that of the General Assembly enhanced. Almost every type of decision, that, by the Charter, appeared to require the concurrence of the five permanent members of the Security Council including the Soviet Union, was taken, on occasion, against the Soviet Union's opposition. This disregard of the constitutional limits on majority rule enshrined in the Charter reached its culmination during the Korean War (1950-53). In particular, the scope of the General Assembly was enhanced, by the Assembly itself, through the Uniting for Peace resolution.

In Western Europe, enthusiasm for majorities began to decline as questions of USA decolonisation, and the economic demands of ex-colonies, became more prominent. For Britain and especially its Conservative government of the day, and France, this process was sharply accelerated by the organisation's role in the Suez crisis of 1956. The conversion of the United States took longer. The issue of whether all UN members should pay an assessed share of the Congo and Suez operations, under threat of eventually losing their General Assembly vote, if they did not, dominated Assembly discussions in the early 60's. The US until 1965, argued that they should; in other words, that the majority could impose obligations on the minority. When, in August 1965, the new American Ambassador to the UN, Arthur Goldberg, finally announced that his country would drop the issue, he did not say that it no longer believed in this principle; but it would not be surprising to learn that he, or his superiors, had reflected that such a principle was no longer likely to serve US interests.

The signs were already there, in the 1960 Declaration on the Granting of Independence to Colonial Countries and Peoples, and the establishment, in 1964, of the United Nations Conference on Trade and Development (UNCTAD) and the birth of the Group of 77; and though Dean Rusk, in 1964, had refused to share the dread of the Assembly's "swirling majorities",(2) American disenchantment with majority rule became unequivocal in 1971, when it signally failed to ensure that the admission of Peking to the Chinese seat did not mean the expulsion of Taiwan. Western, and above all, American, exploitation of the majorities they enjoyed in the General Assembly in these early years is of more than historical interest. It shows how easy it is for a majority rule to override all the obstacles to it a constitutional instrument like the UN Charter apparently provides. For the West, as for the Soviet Union, by the time the General Assembly's Sea-bed Committee first began to discuss the structure of a possible Authority, the dangers of majority voting in international organisations had become manifest. Both at all costs wanted to avoid creating a body in which the main power was given to, or could de facto be exercised by, an organ modelled on the General Assembly, or one freely elected by, and likely to reflect the distribution of voting power within, such an organ. They tried to do this in several ways: by making it more difficult for minorities to be overridden in the plenary organ, by conferring important powers on other organs, differently constituted and with different voting systems (as the authors of the UN Charter had tried to do through the Security and Trusteeship Councils); and by a form of judicial review, through a Tribunal or some such organ, of the constitutionality of all actions taken by the Authority.

Dispute settlement will be dealt with in the next chapter. Otherwise, the debate about structure was mainly focused on the role, composition, and voting system of the executive organ of the Authority, usually spoken of as the Council, although there were also some differences over how the Enterprise should be constituted and how it should fit into the Authority's structure. As to the Council, its existence was never in dispute. What was at issue was the extent to which it should be separately constituted and independent of the plenary organ and given the exclusive right to exercise certain functions. Even if it were, the addition of a provision for judicial review would be necessary to avoid the kind of encroachment by the Assembly on the domain of other organs that occurred in the early years of the United Nations.

The eleven proposals laid before the Sea-bed Committee in 1970 and 1971 (3) differed widely in the extent they would have used the Council to modify the principle of majority rule. The French, which assigned international institutions a negligible role in sea-bed exploitation, said nothing about how they would make what paltry decisions were left to them. Two others, in contrast, (4) made it virtually an executive committee of an Assembly modelled on the General Assembly, with each organ deciding questions by a two-thirds majority. The two from Eastern Europe (5) would have made it consist of five states each from Africa, Asia, Latin America, the Socialist states of Eastern Europe, and "Western Europe and others"; the Soviet Union would have added to these twenty-five members five more consisting of one land-locked country from each of these regions. Both would have given at least each region a veto, at least on some issues. Poland's showed greater flexibility suggesting that different matters might require varying majorities.

Afghanistan and others, (6) reflecting the interests of the "land-locked and geographically-disadvantaged", wanted each state, on joining the organisation, to declare itself "coastal" or "principally non-coastal" and proposed that the Council should consist of equal numbers of each. This would have over-represented the latter, of whom there were never more than a bare blocking third, but it was not without precedent, (7) for instance, in the United Nations Trusteeship Council.

The five remaining proposals all made provision for some designated members of the Council. The UK (8) spoke of "industrialised countries" as good candidates. Canada (9) said no more than that the Council should contain "a proper balance of national interests". Japan (10) proposed a Council of twenty-four of which six would be designated, and of the remaining eighteen, twelve must be developing countries and three "shelf-locked or land-locked". Since, like Canada, Japan proposed that decisions would require only a "two-thirds majority", it would have allowed the designated members, even if united, to be outvoted. This eventuality was eliminated in the United States draft treaty, which, provided for the same numbers of both "designated" and "elected" members as Japan, but would have required majorities in both categories for decisions, so that the designated members - defined as "those six Contracting Parties which are both developed States and have the highest gross national product" (Appendix E.1.) would have exercised a collective veto. Malta (11) proposed three categories of membership. All coastal states with more than ninety million people, and any other coastal states possessing six out of a list of nine qualifications, fell into Category A, and were assured of membership. To these were to be added an equal number of other coastal states, (Category B) and five land-locked states (Category C). Council decisions required both an overall majority and a majority in two out of the three categories. Many of these proposals also provided for more specialised organs. The most important of these was the Enterprise envisaged in the Latin American draft, which became a central element in the Authority's structure, embodying the Group of 77's insistence that the Authority should itself be able to engage in ocean mining. The three Commissions envisaged in the US Draft Treaty (the Operations Commission; the Rules and Recommended Practices Commission and the International Sea-bed Boundary Commission) were to be elected by the Council and each given specific functions, thus further insulating the Authority's decisions from the hazards of majority voting.

Following the 1970 Declaration of Principles the structure of the Authority was considered in the First Sub-Committee of the enlarged Sea-bed Committee. Nevertheless, when UNCLOS III opened in 1973, virtually nothing had been agreed.

There had been two further proposals, both from Italy, in 1972 and 1973. (12) By the first, there would have been a Council of 35 members, fifteen of them "designated" (ten by Gross National Product and five "on the basis of their particular role as coastal states"), in which a two-thirds majority would be needed for matters of substance. The second altered the balance between wealth and coastal state role from ten as against five to eight as against seven, added one land-locked state to make sixteen designated members out of 36, and required a three-quarters instead of a two-thirds majority.

In many of these proposals it is not easy to gauge whether the developed states of the West, either on their own, or with the help of the Eastern European socialist states, would enjoy a potential veto over Council decisions, because it would depend on how many seats would accrue to each region out of these elected by the Assembly without restriction. The practice had grown up in the United Nations of such seats being divided among regions by a prearranged gentlemen's agreement, interpreting the phrase "equitable geographical distribution". What this meant for the Council of the Sea-bed Authority was at first left to the imagination but later made explicit.

According to Buzan (1976,p.173), the most pronounced debate on structure in these early years occurred in 1971-2. The trend of discussions then favoured a flexible view of the Authority's role, allowing it to decide to take on new tasks in the light of experience, and there was much discussion as to whether this initiative here would lie with the Council or the Assembly.

UNCLOS III

The principal contribution of the first two sessions of the conference itself to the solution of the question of the Authority's structure was indirect: it consisted of the precedents set by the Rules of Procedure, on which the first, two-week, New York session was supposed to agree, but did not, but which were, triumphantly, completed, to meet their second deadline, in the first week of the second session at Caracas.

It was generally agreed that the precedents set by UNCLOS I and II, which required only a two-thirds majority to adopt conventions, were not appropriate to UNCLOS III. The rules therefore provided that, before taking a vote, every effort had to be made to reach consensus.

To ensure this either the president, at his discretion, or any fifteen countries, could invoke a ten-day period of delay; and, on the expiry of this period, the conference, before voting on the original question, had to determine that all efforts to reach consensus had been exhausted, and to do this, as well as to take the substantive decision, a two-thirds majority, _including a majority of the states participating in that session_, was required. These provisions suggested a way in which, in a body where all states are represented and count equally, majority rule might be so tempered as to provide some assurance to minorities that they will not be lightly disregarded.

At Caracas (the first substantive session), references to the structure of the Authority were almost always in polarised terms. Developing countries spoke of "democratic" procedures, meaning that the Council should be elected by, and responsible to, the Assembly; developed states of the division of functions between these organs and of the Council's composition needing to be balanced. Among the few exceptions to this trend were Iran and Kenya, the latter suggesting designating six seats, in a Council of 48, for members most advanced in ocean technology. (13) In private, though, many influential developing country delegations at Caracas conceded that the question was not _whether_ the Council should have a decisive say in some questions, and should contain representatives of "interests" as well as members elected by the Assembly, but how this should be done.

The Geneva discussions, which began at the First Committee's twentieth meeting on April 25th 1975, confirmed this. They also revealed two other facts: first, that the "land-locked and geographically-disadvantaged" were demanding more seats on the Council than the coastal states had any intention of conceding to them; and secondly, and less patently, that the USSR and its Eastern European allies, tardy as they had been in accepting the principle of the "common heritage" and the need for a Sea-bed Authority, were now moving much faster than the major Western states towards acceptance of the Group of 77's views on the Authority's structure as well as on the system of exploitation.

The gap was also narrowing between the Group of 77 and the mining states of the West. Alvaro de Soto, the Peruvian spokesman for the 77 on First Committee matters, came close to endorsing what Kenya had suggested at Caracas, a Council which should be a "permanent executive organ", in which account should be taken of "concerned and affected interests" (having in mind, presumably, ocean miners and land-based producers). Leigh Ratiner, for the USA, laid down twelve critical features of the Authority's structure. He insisted that the Authority's powers with respect to exploration and exploitation would fall exclusively within the Council's domain; but conceded the Assembly could also wield "carefully-defined" general policy-making powers. In contrast to these signs of rapprochement, some developing countries, joined by China, Romania and Australia, insisted that the Council be controlled by the Assembly.

The land-locked and geographically-disadvantaged states, including those Eastern European socialist states falling into this category, asked for two things in the First Committee: priority in the distribution of the Authority's revenues: and two-fifths of the Council's seats. This was less than Afghanistan and her co-sponsors had asked for a few years earlier but still higher than their numbers warranted.

On Council composition and voting, the socialist states did not speak with a united voice. Romania's extreme support for majority rule has been noted. Several, including Bulgaria, emphasised the need for consensus; but Poland merely characterised a simple or two-thirds majority as offering inadequate protection to minorities, and East Germany endorsed both the Polish statement and that of Kenya at the previous session.

200

Out of these hints of accommodation, Chris Pinto, who as chairman of Committee I's Working Group was actually conducting the negotiations, was able to produce a fairly acceptable text in this respect. The Assembly would be "the supreme policy making organ" but would <u>not</u> be able to impinge on the Council's prescribed functions. The latter would have thirty-six members, half representing specific interests, equally divided between developed and developing countries; the other half would be divided among the regions; and decisions would have required a three-quarters majority. This proposal would have ensured that the Western industrialised states would jointly wield at least the ten votes needed to block a decision. No subsequent chairman's text offered them as good a deal.

However Pinto's text, as we have seen, was unofficial, and Engo, in writing Part I of the SNT, made significant changes in it. Only twelve Council members were to represent "interests"; of the six assigned to industrialised countries, one was earmarked for Eastern Europe: and a majority of two-thirds plus one was sufficient for Council decisions. Depending on how the twenty-four "elective" seats were assigned to regions, this could have given the Group of 77 an overall majority in the Council. At best, the Western states would have needed communist votes to block a decision on which the Group of 77 were agreed.

The Pinto text would also have made it more difficult for a vote to be taken. In the spirit of the conference's rules of procedure, the Council would first have had to decide, also by a three-quarters majority, that "reasonable" efforts had been made to reach general agreement. The same principle, mutatis mutandis, was to apply to decisions of the Assembly which were to require a two-thirds majority, and a minority making up a quarter of that body's membership could insist on a vote being deferred pending an appeal to the Tribunal on legal grounds. A seemingly similar provision in ENGO's SNT was rendered pointless by requiring a third of the membership to invoke the delay, a fraction virtually sure of being able to defeat the proposal anyway.

Engo's alterations to the Pinto text may well have resulted in a chance of early agreement on the Authority's structure being missed. Certainly the SNT aroused strong feelings, not least in its structural aspects. The American amendments, conveyed privately to Engo in December 1975, called for drastic changes. Out of a Council still of thirty-six only twelve would have been elected, with at least two from each of the five regions; the remaining twenty-four were to consist of six industrialised countries; six representing special categories of developing countries; six of the main producers of the minerals in question; and six of the main consumers. Decisions were to require a three-fourths majority of those present and voting, a simple majority in at least three of the five categories specified, and, prior to taking a vote, a determination by a similar combination of majorities that all efforts to reach agreement had been exhausted. The requirements of a simple majority in each of a certain number of subdivisions was repeated with respect to the Assembly, where members were divided into five categories ("industrialised", "main producer", "main consumer", "land-locked and geographically-disadvantaged", and "others"), and a decision required both a two-thirds majority of those present and voting, and a simple majority in each of four out of the five categories. There was also a requirement, similar to that proposed for the Council, that, before voting, the Assembly should determine that all efforts to reach agreement had been exhausted; and, as in the Pinto text, it was to be open to any group, comprising a quarter of the membership, to delay a vote pending an appeal on legal grounds to the Tribunal.

Some intersessional meetings at New York in February gave Engo an opportunity to respond to these amendments, which would have destroyed the voting power of the Group of 77, even in the Assembly. In PBE6, of February 11th, 1976, while in no way endorsing the division of the Assembly into "categories" for voting purpose, he did introduce some fresh constraints on the power of the majority. First, unless hatched by consensus, a decision on an "important question of substance" would not come into effect until two months after the end of the session where it was adopted, and then only if one-third of the membership did not notify the Authority's Secretary-General of their objections to it. Secondly, following the example of the Rules of Procedure of the conference, he proposed that either the president, at his own discretion, or any fifteen members, could delay a vote being taken for one period of five days, to increase the chances of reaching consensus. Finally he adopted a feature common to the US Amendments and the Pinto text, that the fraction of the membership that could delay a vote, by appealing to a tribunal on the legality of the matter, should be one-quarter rather than one-third.

After the fourth session in the spring of 1976 these proposals were incorporated into the Revised Single Negotiating Text. There was no change in that Text in respect of the composition and functions of the Council, but it was made clear that further negotiations would be needed on them, expected to take place at the fifth session fixed for later that year.

The fifth session proved too divided over other aspects of the RSNT to permit consideration of the voting and composition of the Council, but at the end of the session both the USA and the USSR privately circulated new amendments on these points. For the first time in these negotiations, and possibly in the history of international organisation, the Soviet Union was more hospitable than the USA to the principle of majority voting. Ratiner, for the latter, withdrew his own earlier proposals on the embarrassing grounds that they were not now believed to be adequate to protect US interests. The Council's size and composition would have been as in the 1975 book of US amendments, except that nine seats would have been reserved to land-locked and geographically-disadvantaged states, and that the membership would have had to include those accounting for half the value of both the total production, and the total consumption, of the relevant minerals. Voting was to be weighted according to shares of these totals, and valid decisions required not just a three-quarters majority of those present and voting, but the support of states accounting for half the total production of the membership and half the consumption. The adoption of the budget required the support of states accounting for three-quarters of these totals and might, therefore, have been logically impossible in a validly-elected Council. Moreover, such weighting would seem to entail that whenever a miner extracted the smallest quantity of a new metal, the weights would change. If ever a proposal was unworkable, this was.

The Soviet proposal, by contrast, was straightforward. There were to be six categories: those elected by regions (12, with at least two from each); "advanced" - i.e. ocean mining states (6); special developing country interests (6); consumers (4); producers (4); and the LLGDS (4). Eastern Europe would have been represented by the "advanced" state, one "consumer", and one of the LLGDS, as well as at least two from the regional seats. Decisions required a two-thirds majority, including a majority in at least three of the categories. The five socialist votes would have been less than half of a blocking minority, but would probably have been needed, to form such a minority, in conjunction with the West, against a united Group of 77-producer alliance.

The intersessionals of February 1977, and the early meetings of the Chairman's Working Group of the First Committee at the sixth session, both led by the Norwegian Jens Evensen, did not discuss structure at all. While Evensen was away for a week, Engo resumed the chairmanship and launched the Committee into a desultory discussion of the subject of the Council; but interest revived only with the production by Evensen of a series of "compromise formulae" on this and other First Committee matters. In essence for the Council he followed the recent Soviet and 1975 US proposals for "chambered" composition and voting. The five "chambers" were to consist of eighteen regionally-elected members, including at least two from each region; six representatives of developing country interests; four "ocean mining" states; four importers of the metals in question; and four exporters (i.e. land-based producers). Decisions were to require a two-thirds majority, including a simple majority in at least four of the five chambers, an ingenious arrangement which would have permitted consumers to ally, to block a Council decision, with producers against monopolistic practices by ocean miners, as well as, more obviously, with ocean miners themselves, against developing countries and producers.

For the Informal Consolidated Negotiating Text, Engo kept Evensen's provisions relating to the functions and voting system of the Assembly, but significantly altered the composition and voting of the Council. The regions would still have elected half of its thirty-six members, but each region would have been assured of only one seat. Of the "exporting" group, at least two would have had to be developing countries; and one of the "importers", as well as one of the "technologically-advanced", (as in Evensen) was to come from Eastern Europe. But instead of requiring majorities in a specified number of "chambers", decisions on substance were simply to require a three-quarters majority overall.

The Council as proposed in the ICNT was very similar to that of Pinto's 1975 "shadow" text, with the important difference that, now being assured of only one seat as a region, the West could not muster the ten votes needed to block a Group of 77 proposal, without support either from the Eastern Europeans or from any developed countries among the producers. This was enough to make it unacceptable to the USA and most, if not all, prospective mining countries, in that, in the words of Elliot Richardson's press statement of 20 July 1977, it would

> fail adequately to protect minority interest in its system of governance and would, accordingly, threaten to allow the abuse of power by an anomalous majority."

When at the seventh session, at Geneva in the spring of 1978, it was decided, in response to criticisms of the way in which Engo had altered the Evensen Text for inclusion in the ICNT, to split the remaining issues before that committee among three negotiating groups, it was perhaps indicative of a sense that this would be one of the less difficult areas of negotiation that questions of the Authority's structure were assigned to the group which Engo chaired, NG 3.

NG 3, however, made little progress at Geneva. Some minor changes were made, commanding general support, but Engo had little sympathy for any form of chambered voting which he equated with 'collective vetoes' and 'weighted voting' (NG 3/2 p.6). saw increase in the representation of the West as possible only if the Council were to be enlarged, and claimed to detect a general feeling that a two-thirds majority rather than a three-quarters majority would be enough.

There was thus some excuse for the apparently extravagant complaint of Elliot Richardson after the session, that:

"The differences remain clear-cut and fundamental. The developing countries tend to regard the Council as simply an arm of the Assembly with priority for the principle of equitable geographic distribution" (Richardson, May 22, 1978, p.6).

This would imply that UNCLOS III had so far made no progress in devising a structure that would protect minorities from majorities. Frustrating as the last year had been for the developed states, this was exaggeration. The principle of the special representation of interests, including those of the sea-bed mining states and of the sea-bed mineral importing states, was not disputed; and it was agreed that "interest" states should have parity with "elected" states in the Council overall. A likelier explanation of the frustration is that states were unwilling to concede anything about machinery until issues about the system of exploitation had been settled. Engo was probably right in saying "I firmly believe that each side to the negotiating battle does not wish to give in or accept compromise until most of the package of core issues in the First Committee mandate is considered together". (14) Events at any rate bore out this prediction.

The voting system of the Council was the last major issue to be hammered out in Richardson's time. When it was, in the summer of 1980, the collegium was sufficiently proud of the achievement, and satisfied by the responses of delegations to it, to describe the new (third) revision of the ICNT, which then emerged, as the "Draft Convention".

For some time the voting question was seen as that of specifying a number of states, lower than the ten required by the ICNT, whose combined opposition would be enough to defeat a proposal in certain "particularly sensitive" areas. NG3, and the newly established Group of 21, considered replacing ten by anything down to five, but without reaching agreement on any figure.

Between 1979 and 1980 Tommy Koh, successful with the financial questions of NG2, was brought into the negotiations on Council voting. Progress was made by means of a very small group consisting at first only of the USA and six members of the Group of 77, (15) to which were added first the EEC and Japan, and then the USSR. The president also met daily with fifteen representatives of the Group of 77. There was some chagrin among those excluded but the Group of 77 stuck by the agreement made by its representatives. The agreement that emerged divided issues of substance not just into the more sensitive and the less, but into three categories, requiring, respectively, a two-thirds majority of those present and voting, a three-quarters majority of those present and voting (which in each case had to constitute a simple majority of the Council's membership); and consensus.

Consensus meant in effect that _every_ member of the Council had a veto, though it was hedged by a provision whereby, if there was a prospect of its use, the president would constitute a Conciliation Committee within three days, which would report to the Council within a further fourteen. The Committee would try to reconcile the differences. If it was able to produce a generally-acceptable proposal it would do so. If not, it would set out the grounds on which a proposal was being opposed. (16)

For the issues thus seen (but not officially characterised) as most sensitive, a curious compromise had thus emerged. In 1979, there was argument as to how big the blocking minority should be, within a range of from five to ten. In 1980, agreement was reached that it should be one! One explanation is that the USSR was willing to be outvoted by a majority only if the West needed its support to block a decision, and the West only if they did not, which entailed that agreement was possible only if neither could be outvoted. For this to happen, however, the Group of 77 had to abandon, at a stroke, its voting dominance over such issues.

By the new accord, kept in all subsequent texts, five categories of decision required consensus: measures to protect land-based producers against adverse economic effects, on the recommendation of the Economic Planning Commission (Article 162, 2(d)); recommendations to the Assembly of "rules regulations and procedures" to govern benefit sharing (Article 162,2 (a)(i)); the adoption and provisional application, pending approval by the Assembly, of rules regulations and procedures governing activities in the Area and "the financial management and internal administration of the Authority" (Article 162,2(n)(ii)), amendments to Part XI (other than those arising from the review conference); and the _disapproval_ of plans of work recommended for approval by the Legal and Technical Commission, except that in considering such plans of work a sponsoring party could not veto disapproval of the plan of work it sponsors, that is, if there is a proposal to disapprove a plan of work, and the only objection to the disapproval came from the plan's sponsor, the Council would disapprove it.

These five categories covered a very large range of ground. For the developed states, the worry that mining would be unable to begin because _no_ rules had commanded consensus in the Council was allayed by the presumption explicitly articulated by Richardson, in his speech of August 26th as a condition of accepting the compromise, that the initial rules would be arrived at in the Preparatory Commission and thus precede decisions on ratification. Thereafter, they could be amended only with the consent of each member of the Council at any given time. (Assembly approval was not vital, since in its absence they could still be applied "provisionally").

Insistence that consensus should also apply to the adoption of rules governing revenue-sharing, a question which in earlier texts had been seen as safe enough to be entrusted to the Assembly, was a reaction to the Group of 77's insistence that the beneficiaries could include not just parties to the convention but also

"peoples who have not yet attained full independence or other self-governing status recognised by the United Nations in accordance with General Assembly resolution 1514 (XV) and other relevant General Assembly resolutions" (Article 140)

which was interpreted as covering the PLO.

In the meetings of Committee I, and subsequently of the plenary, with which the ninth session closed, only the smaller industrialised powers complained at the new proposals. Unless they qualified as "consumers", "producers" or "ocean miners", they would have to queue for the one regional seat allocated to "Western European and others". The superpowers, and their major allies, were adamant that the question of the Council's size and composition, which had remained essentially unaltered since the ICNT, could not now be reopened.

The structure of the Authority, including the question of its composition, was however soon to be raised by the critics of the draft convention inside and outside the new Reagan Administration. Much play was made of the fact that, while the Eastern European socialist states were explicitly assured of three seats in the Council, the USA was not specifically assured of one. For instance, in his address to the resumed tenth session, at Geneva, on August 3 1981, the US delegation leader, James L. Malone, mentioned as one of the points causing difficulty for his country that

"it was uncertain that the United States and other technologically-advanced Western countries would be appropriately represented on the Council of the proposed... Authority. Also, there was concern as to the ease with which the voting system (requiring consensus for all major decisions) could be used to paralyse the Council" .(17)

The "Approaches" paper, produced before the eleventh session had begun in February 1982, listed three kinds of objection to the Authority's structure (other than that of the Enterprise) as provided for in the text formalised the previous year (L78). These related to: the Council's composition; its voting rules; and the role of the Assembly. On composition, assured representation was sought for the USA and the major industrialised countries; on decision-making, two complaints were made. The three-quarters majority requirement, in itself, was seen as too permissive; the consensus rule, by contrast, as too restrictive. The US wanted the Western industrialised states, in combination, to be able both to block more decisions, and get more decisions carried against the opposition of others. Of the two, the latter requirement was seen as the more vital. As to the Assembly, the US wanted to change the wording of L78 so that it would no longer be designated "the supreme organ" or have residual powers, and, where required to act "on the Council's recommendation" would be explicitly confined to approving that recommendation or referring it back, with comments. This latter restriction might perhaps have been regarded as already implied by the wording of L78. More fundamentally, though, objections were raised to the Assembly's own decision-making process and the difficulty of delaying a vote in order to achieve consensus. The Green Book of US amendments, issued in March, proposed major changes in all three areas. On Council composition the two categories of "investors in ocean mining" and "largest importers" were kept only in name. Their six seats (other than the two Eastern European representatives) were to be occupied by six of the seven UN members with the largest assessed contributions to its budget. Since the USSR is in fact the second highest contributor, this would have left no room for choice for the other six. There were to be six different modes of taking decisions in the Council. Three of them corresponded to those of L78 - simple majorities (for procedural question); two-thirds majorities, including a majority of the Council's members, for the less contentious matters of substance, to which the Green Book added some hitherto requiring a three-quarters majority; and "consensus", which was now to apply only to amendments to Part XI. The three-quarters majority, however, was made much more exacting. It was required to include a majority of each of the Council's five categories, two of which would have, in their composition, no distinguishing feature. Thus if two of the six largest contributors to the UN (other than the USSR) opposed a proposal supported by everyone else it would fail if they happened to be assigned to the same category but succeed if they were in different ones. Such illogicality testified to the Green Book's hasty drafting. The two other modes of voting were to be applied to the adoption of the Authority's "rules, regulations and procedures", if the Council's Legal and Technical Commission, which would consist of one nominee of each member of the Council, failed to reach consensus on them within a year of its first meeting. With respect to benefit-sharing, rules were to be adopted by a majority of the five developing country Council members which were the largest producers, on land, of nodule minerals; in respect of activities of the Area, by a majority of the seven largest contributors to the UN budget. Thus in both these cases, when it came to a showdown, most Council members would have had no say.

Similarly drastic surgery was envisaged for the Assembly. In addition to verbal changes along the lines indicated earlier, its voting system was to be transformed so as to emasculate the Group of 77's majority in it. Decisions would have required not only a three-quarters majority of those present and voting, and a simple majority of all members, but also a three-quarters majority of those that were members of the Council. These proposed changes in the Authority's structure explain how Leigh Ratiner was able to speak of the US having "demanded virtually autocratic ruling powers over the Sea-bed Authority" (Ratiner 1982, p.1013).

Nevertheless, the US secured support among the six other major mining states of the West (Belgium, France, Italy, Japan, the UK and West Germany) for formal amendments going quite far in that direction. In L121 the seven proposed that the Council include six of the eight most heavily assessed UN contributors. For Belgium, this was peculiarly self-sacrificing. It would have reduced her chances of a seat on the Council as an "ocean mining" or "consumer" state. Her only other hope of access would have been as the sole regional representative of "Western Europe and others", for which honour competition, as we have seen, would have been extremely keen.

These proposals specified four modes of Council decision-making on questions of substance: two-thirds majority, three-quarters majority (in each case including a simple majority of those participating), a _qualified_ three-quarters majority, and consensus. The qualified three-quarters majority, which was to apply to many categories of decisions, including some for which the draft convention had prescribed only a two-thirds majority, imposed even more stringent conditions than the corresponding proposal in the Green Book, namely a majority in each of the five "regions", as well as in each category of interests. Whether the members representing "interests" would regroup into regions for this purpose, or the sole regional representative of the Eastern European and Western European regions would each have enjoyed vetoes, is not made clear.

The idea that rules might be ultimately adopted by a majority of the seven largest UN contributors or (relating to benefit-sharing) by the five leading developing land-based producers on the Council, was dropped. The adoption of both kinds of rules remained subject to consensus. The Assembly's role and voting system also remained unchanged, apart from a clarification of the application to it of the "separation of powers" principle (Article 158).

Thus, paradoxically, the Council's potential for paralysis, which had been the main target of US criticism, would have not merely remained, but have been enhanced, if the seven power amendments of which it was a sponsor had been accepted.

Attempts were made to bridge the gap between the seven and the Group of 77, notably by a Group of 11 "friends of the conference". The proposals, not without some advantage to their proponents, would have assured "Western Europe and Others" (WEO) of two of the "producer" seats, and would have created a category of decisions for which a majority of three-quarters plus one of those present and voting was required, thus giving WEO a collective veto on its own. The prospects of success for the Group of 11's batch of compromise proposals were, as we have seen, killed by the strong reservations expressed by the USA.

Both these sets of amendments were withdrawn, in return for the withdrawal of others. The president was able to report consensus on two changes in the Authority's structure, which were adopted on April 29th: a seat was reserved for the "world's largest consumer" of nodule minerals, that is the USA, which merely made explicit what had always hitherto been assumed; and regulation of non-nodule minerals from the Area was made to require a consensus decision of the Council, rather than an amendment of the convention. Since, in the specific case of amendments arising out of the review conference, a three-fourths majority could, in the end, adopt amendments and by their ratifications, bring them into force for the rest, this change would actually have made regulation of such mining more difficult. (18)

Mention was made earlier (chapter 7 above) of the claim in the US "Approaches" document that "the real power to grant or deny access rests in the Legal and Technical Commission". This claim must be qualified in the light of the very stringently defined conditions, set out in Annex III (Basic Conditions of Exploitation Articles 4 and 6, under which the Authority could deny an applicant a contract, and even if it did, there would be a right of appeal. Nevertheless, though the commission's discretion, if any, is, by the convention, very limited, it is worth examining where it fits into the Authority's structure.

The convention provides for two commissions subordinate to the Council, the other being the Economic Planning Commission, and Article 163, specifying their structure, applies to both of them. Each is to have fifteen members, though the Council, by a three-quarters majority, can decide to increase the size of either. They are to be elected by the Council, by a similar majority, from among nominations by States Parties, no one state nominating more than one member of each commission. There is a provision that 'due account must be taken of equitable geographical distribution

and the representation of special interests' (Article
163.4), which is probably more than the platitude it seems,
since on the loss of any member through death, incapacity or
resignation, the replacement must be from the same region or
interest. The mode of decision-making of each commission is
to be established by the Council's "rules, regulations and
procedures", which means, of course, by consensus. This was
decided when the Council's own decision making was agreed
upon; before the third revision of the ICNT the commissions
were to take decisions by majority vote. Two other
additions were made at that time: the requirement that for
both bodies members should have "the highest standards of
competence and integrity with qualifications in relevant
fields", and specifically for the Legal and Technical
Commission, that its recommendations with respect to written
plans of work (which include the approval of contracts)
should be "solely on the grounds stated in Annex III" and be
"fully reported" (and thus presumably, explained in terms of
that annex). This commission would have several other
functions, including that of recommending, to the Council,
the disapproval of areas or the issuance of an emergency
"stop-work" orders; but these powers to impede exploitation
would require a further positive (3/4 majority) vote of the
Council; only on the approval of contracts is the
commission's vote so crucial. Clearly the intention is to
make it a purely technical body. The possibility that its
recommendations on this would operate to deny access by
coming to reflect the political views of the preponderance
of the Council's membership, though it cannot be
theoretically ruled out, would require that explicit and
precise rules laid down in the convention were blatantly
disregarded both by the commission itself and by the Seabed
Disputes Chamber of the Law of the Sea Tribunal.

Leigh Ratiner hints that in spite of the radical nature of
the amendments formally proposed by the seven mining states,
the fundamental cleavage between the USA and the Group of 77
was not over the Authority's structure. That structure,
particularly as it comcerned the Council and its
commissions, and the Assembly, was not ideal; but the
meeting of minds that seemed to have occurred in 1980 was
too fresh in the memory of the Group of 77, and had required
of them too many concessions for them to be hospitable to
the drastic demands generated by the Reagan review.

The Enterprise

Little consideration was given in the early years of
conference to the structure of the Enterprise. It was
widely thought of as a means by which developing countries
could themselves participate in, and through their numerical
strength, control, the mining of the Area.

An early draft of its statute, which appeared in the RSNT, required the Assembly to elect its Governing Board, of thirty-six state representatives, by the same criteria as applied to the Council (at that time "mining and importing" states, six "special interest" developing countries, and twenty-four elected by regions). The Governing Board would elect the Director-General. Notwithstanding this the Enterprise was (and remained in all drafts) subject to the Council's 'policy directives and control' (Annex II, para 2(a)).

When in the ICNT the composition of the Council gave it a somewhat different complexion from the Assembly's, the latter was given more power over the Enterprise, which was also to be subject to its "general policies". The Governing Board, now reduced to fifteen, was to be elected by the Assembly, essentially according to the "equitable geographical distribution" criterion. The Director-General, though, was to be elected not by the Governing Board but by the Assembly on the recommendation of the Council.

As the rich states committed themselves to subscribe more and more capital to put the Enterprise more promptly into business, they not unnaturally demanded more control over its operation. In the first revision of the ICNT, this was made subject to all the Authority's "rules regulations and procedures" unless otherwise specified. A year later, in the second revision, many changes were made which remained in the convention as adopted; the formula of "election by the Assembly on the recommendation of the Council" was applied to both the Governing Board and the Director-General, the latter also needing the nomination of the former (Annex IV Article 5 and Article 160, para 2(c) of the main text).

Since these three bodies thus had to agree on a name, there was a not inconsiderable prospect of protracted wrangle and delays.

There was some debate over the Enterprise's "autonomy". It continued to be subject to the directives of the Council and the Assembly's general policies, and was given, against American opposition, the right to negotiate tax exemption with a host state (Annex IV, Article 13.5) (given its uniqueness, it could hardly fail, in practice, to enjoy, like multinationals, this kind of bargaining power). To preclude its acquiring further competitive advantage it was required, after a grace period, to make payments to the Authority on the same basis as private and state contractors (Article 10). No objection was raised to making explicit, what was already understood, that it could engage in transporting, processing and marketing nodules as well as mining them.

The Reagan review led to demands for two kinds of changes concerning the Enterprise: changes in the resources at its disposal and changes in its structure. As explained in chapter 7, by the US amendments, contractors would no longer have to sell technology to the Enterprise (unless they also sold it to third parties); and "reserved sites" were to revert to the "private and state" sector (with very much emphasis on the "private") if unused over a specified period. It was also proposed that the Enterprise should have to pay contractors for any training of its personnel. Structurally, the Governing Board was to be required, so long as the Enterprise was in debt, to include in its fifteen members nominees of creditors accounting for at least 50% of that debt. Since, in terms of UN assessments, this could amount to only four members, and since the Board took its decisions by simple unweighted majority, this would not have given the creditors control of it. Finally, the US also sought to free the Enterprise from control by either the Council or the Assembly (except in respect of compliance with rules, regulations and procedures). The seven-power amendments followed the US lead on transfer of technology and the reversion of "reserved" sites, and on giving major creditors seats on the Governing Board. Neither the US nor the seven-power amendments would have prevented the Enterprise from flourishing; it was still, after all, to be equipped with the capital necessary to exploit one site.

The proposals emanating from the Reagan review, then, would have weakened the Enterprise, but by no means have destroyed it as a significant global actor; in contrast to those directed at the rest of the Authority's structure, which were extreme and illogical. In both cases, their timing virtually ensured that they failed, which perhaps they were intended to.

NOTES

1. UN Charter, Article 27.3 and Article 108. The point is elaborated in Ogley (1972).

2. Dag Hammarskjold Lecture, January 10th 1964, cited in Russell January 1968, p.356.

3. see above, Chapter 7, n.5 for a list of these.

4. A/AC 138/33 (Tanzania) and A/AC 138/49 (Chile and others).

5. Those of the Soviet Union (A/AC 138/43) and Poland (A/AC 138/44).

6. A/AC 138/55. Four of the seven were European states.

7. The Trusteeship Council, one of the six principal organs of the United Nations, was so constituted as to ensure an

equal division between states administering Trust
Territories and non-administering members, although a
further requirement that all permanent members of the
Security Council also be permanently in the Trusteeship
Council interfered with the balance as the number of
administering powers declined; the example was followed
even in a committee set up by the General Assembly, shortly
after its inception, in order to subject information from
non-self governing territories to a not dissimilar scrutiny.
The rationale for applying this principle to the Authority's
Council depended on whether it would have any say in
determining the limits of the area it administered. In any
such decision about limits, a balance between "coastal" and
"primarily non-coastal" interests might have had a lot to be
said for it. In other decisions of the Authority it is hard
to imagine divergences of interest following these lines.

8. A/AC 138/46.

9. A/AC 138/25.

10. A/AC 138/63.

11. A/AC 138/63. The Maltese plan, as explained earlier in
chapter 7, was designed not for a Sea-bed Authority as such
but for a set of institutions to manage "ocean space"
generally, and was called a "Draft Ocean Space Treaty."

12. A/AC 138 SC.1/L15(1972) and A/AC 138 SC.1 L24(1973).

13. Adede's statement in C.1., July 16th 1974 (UNCLOS III,
Records, Vol II 1975, pp.19-20); but these were very much
the minority among developing countries. The contrary view
was so regularly expressed that Pakistan ignored the strong
(if politely worded) reservations of certain developed
countries, and went so far as to claim that a consensus was
emerging that the Council should be subordinate to the
Assembly and elected by it, and have no permanent members.

14. NG 3/2, p.6. See UNCLOS III, Records, Vol X 1978,
p.79.

15. Believed to be Brazil, India, Jamaica, Peru, Singapore
and Tanzania; see Commonwealth Secretariat 1982 p.58.

16. ICNT Rev 3, Article 161.7(e).

17. United Nations Press Release BR/81/20 31 August 1981,
p.6.

18. By L78, the procedure for amending Part XI of the
convention, as a result of the review conference, permitted
amendments to be adopted, in the absence of agreement, by a
two-thirds majority, and to enter into force for all parties
when ratified by the two-thirds of them. At the President's
suggestion, this was changed to three-quarters.

10 Dispute Settlement

In the context of the attempt, since 1970, to create a new, generally accepted, law of the sea, in which an international sea-bed authority would play an important role, the idea of "dispute settlement" was seen as having two applications.

If there was to be an active Authority, making rules, entering into contracts, and issuing orders to contractors, it was seen in the West, at any rate, as essential that an aggrieved contractor or applicant, whether or not it was a state, should be able to appeal against the Authority's decisions. At the same time, given the enormous range of topics, besides sea-bed mining, with which the convention was to deal, some dispute settlement mechanism was widely considered, especially by the maritime powers, as at least highly desirable. This is well illustrated by a remark of Leigh Ratiner (chief US representative in the First Committee of UNCLOS III except during the Carter Administration) at the eighth annual conference of the Law of the Sea Institute, then located at Kingston, Rhode Island, in June 1973, that

> "the treaty will have many diverse rules with lots of ambiguities. Is there going to be a useful law of the sea treaty that settles all these issues, if in fact it does not have a section for compulsory settlement of disputes?" (Gamble and Pontecorvo 1974, p.16).

Because of this twofold application of the idea, UNCLOS III, in considering what methods of dispute settlement should be devised for the system of exploitation of the international area, has tended to oscillate between treating the question in relation to all disputes that might spring from the convention - and thus dealing with it in what was, in effect, a fourth main committee, called the "informal plenary" and chaired by the conference president - and treating it as an integral part of the regime and machinery for the Area, and thus falling within the domain of Committee I.

In essence, the problem gave rise to two broad sets of issues. First, what kinds of decisions could be appealed against, and by whom, and on what grounds; secondly, what kind of body or bodies would be given the job of deciding such appeals, and if they did not exist already, how would they be constituted, and their members appointed?

The importance of dispute settlement was signalled in the US Draft Treaty of August 3 1970, which called for a Tribunal which could hear any inter-state dispute about the sea-bed; challenges, on any of five grounds ("a violation of this Convention, lack of jurisdiction, infringement of important procedural rules, unreasonableness or misuse of powers") to any of the Authority's decisions; and any relevant dispute submitted to it arising out of an agreement, licence or contract.

Such a system would have afforded private (and state) enterprises, operating, or seeking to operate, in the international area, virtually complete protection against any decision that the tribunal considered unwarranted. It is not clear that it would equally have protected the Authority, since complaints by it against operators required the support of an Operations Commission before the Tribunal could have heard them.

Dispute settlement was mentioned, vaguely, in the Declaration of Principles. There was also some provision for a Tribunal in the proposals of the UK, Canada, Japan, Poland, Afghanistan and others, and for an International Maritime Court in Malta's. Chile and its co-sponsors offered dispute settlement as a heading without formulating a text. Tanzania proposed to refer all disputes not settled by the Authority to the International Court of Justice. Only France and the USSR made no provision for it (except by agreement of the parties); as already described, (1) both then saw the Authority's role as minimal. In spite of this fairly widespread acceptance that dispute settlement would have some role in the new sea-bed regime, the Sea-bed Committee, by its demise, had achieved no progress in determining what that role would be.

The subject was not discussed at Caracas, except in an informal private group in which the American international lawyer, Louis B. Sohn, was prominent. Few states mentioned it in the First Committee's 'general debate';(2) and Tanzania was virtually alone in explicitly questioning the necessity for mandatory settlement and advocating a balance between "acceptability" and "viability", on the one hand, and "ability to compel". Even the four proposals as to the basic conditions of exploitation (from the USA; the Group of 77; the eight EEC countries other than Ireland; and Japan) made surprisingly scant reference to how disputes about their application were to be resolved.

From then on, until the end of 1976, the First Committee maintained control of the issue of sea-bed authority dispute settlement; although the "informal plenary", the successor to Sohn's group, issued two SNTs (the first being considered premature in some quarters) and, at the end of the fifth session, an RSNT, it left sea-bed questions to the First Committee; thus we must look to the Engo SNT, the Pinto "Shadow" SNT of 1975, and the (Engo) RSNT of 1976, for developments over these years.

Both the SNTs made elaborate provision for obligatory dispute settlement and included a Tribunal among the principal organs of the Authority, with access for member states and their nationals, and jurisdiction over the interpretation or application of the convention, and of contracts pursuant thereto; further, states (only) could challenge any decision of the Council, its subsidiary organs or the Assembly, on four of the five grounds listed in the US Draft Treaty ("unreasonableness" being omitted). The Pinto text went even further, allowing an applicant for a contract to challenge the Authority's refusal to give him one, and any "interested" individual - i.e. company - to challenge a decision of the Authority affecting him.

It was, however, the official Engo text to which the participants in UNCLOS III needed to respond and, not surprisingly, the list of amendments proposed by the United States to it in December 1975 would have reinstated the points dropped from the Pinto version and would have made some minor additions to it. These the RSNT accepted virtually in total, and with hardly any challenge at the fifth session, fundamentally split though that was on the system of exploitation. It seemed generally agreed that, whatever system was embodied in the treaty and in contracts made pursuant to it, its member states and, where applicable, their companies on the one hand, and the Authority on the other, would have an almost unfettered right to bring each other's violations of that system before a tribunal with binding powers.

The intersessional negotiations of February 1977, called by Jens Evensen of Norway, proved these interpretations deceptive. The Mexican head of delegation, Ambassador Castañeda, in what he described as "a personal and informal contribution to negotiations" thought that some states would be willing to accept a mixed (i.e. "parallel") system, provided always (emphasis in original) that other conditions not contained therein are accepted at the same time." One such condition was "a reasonable limitation on the functions of the Tribunal": it was not, that is, to pronounce on the legality of any provision of a general nature which has been adopted by an organ of the Authority, and which constituted the legal basis for the measure questioned; and the Tribunal's decisions, like those of the International Court of Justice, were not to have established binding precedents. (3)

217

This proposal would seemingly have restored to the Authority, by the back door, much of the discretion that a genuinely parallel system necessarily precluded. The latter meant "assured access" for states and companies. They must know, from the terms of the convention, the broad dimensions of what they had to do to gain, and keep, a contract. In the absence of a tribunal capable of declaring the Authority's "rules, regulations and procedures" illegitimate (at least in their application to a given case), there would be nothing to prevent the Authority making whatever rules it chose.

Such a proposal might then have been expected to run into opposition of a fundamental kind from potential exploiters of the sea-bed and their governments.

Nevertheless, Evensen, in his compromise formulations on the Sea-bed Dispute Settlement System, circulated to the "chairman's working group of the First Committee" on July 5th 1977, included an article (Article 37) embodying the Mexican suggestion, which also appears in the Informal Composite Negotiating Text, with the added proviso that the Tribunal would have no jurisdiction with regard to the exercise by the Authority of discretionary power and could in no case substitute its own discretion for that of the Authority.

Thus by now the dispute settlement system of the Authority had been thoroughly reopened, at the instance of Castañeda and Evensen, without, at the time, appearing to evoke an outcry from the West.

When the outcry came, in Richardson's press statement of July 20th, dispute settlement was not mentioned and it was not among the seven 'hard core' issues, three of them concerned with sea-bed mining, for which separate negotiating groups were established by the seventh session at Geneva in 1978, when it finally got down to business.

Why this was so is difficult to explain, particularly since this transformation in the scope of dispute settlement was accompanied by important changes in its structure. Part I of the RSNT, like its predecessor(s), had provided for a separate Tribunal, as an organ of the Authority, whose judges would have been jointly elected by the Assembly and the Council. The ICNT, instead, followed Part 4 of the RSNT (i.e. that emerging from the informal plenary) in assigning sea-bed disputes to a Sea-bed Disputes Chamber (henceforth referred to as SBDC) of a treaty-wide Law of the Sea Tribunal. Although questions as to "its jurisdiction, powers, and functions, and access to it" were left to be determined by the First Committee,(4) this meant that its judges would be drawn from a panel of twenty-one (the members of the Law of the Sea Tribunal) appointed by a process outside the Authority's control, involving a meeting of all parties to the Treaty (requiring the attendance of two-thirds of them to be quorate) and a vote by a secret ballot, successful candidates requiring a two-thirds majority of the votes cast (a simple majority of all parties). Of the twenty-one, not fewer than three were to be drawn from each of the five geographical groups (Asia, Africa, Eastern Europe, Latin America, and Western Europe and others).

These provisions would have assured the Third World of a minimum of nine of the twenty-one judges; up to six more, if the Group of 77 could have agreed on nominations for the six "floating" places and had voted accordingly. In such a system, nationals of Third World states would almost certainly have constituted a majority.

At the sixth session, the First Committee did not immediately treat Part IV of the RSNT as definitive. Evensen's compromise of June 29 listed a Tribunal among the organs of the Authority, though with a qualifying footnote to the effect that the structure of the Sea-Bed Dispute Settlement System, and thus the form of the body in question, was still to be negotiated. Six days later, in his next paper, he referred to "the Sea-Bed Dispute Chamber of the Law of the Sea Tribunal" and explicitly assumed, in a footnote, that "the institutional and procedural aspects" of such a system would be dealt with in Part IV of the convention, that is within the framework of the Law of the Sea Tribunal. The ICNT followed Evensen in this, except that the job of selecting judges to constitute the former from among the members of the latter was to be done by the Assembly alone, every three years, by the majorities normally required for its decisions of substance.

This was a far cry from the US Draft Convention. By that proposal the Council alone, through its normal concurrent majorities of industrialised and developing members, would have elected the body which was to protect members and their companies against its own decisions and those of its commissions. By 1977, without specific protest by the West, this body had become a chamber selected exclusively by the Assembly from the Tribunal, elected, again, on the principle of one state one vote, in which nationals of Third World countries could reliably be expected to constitute a majority.

It was as if, at this time, dispute settlement by an acceptable organ was not being regarded as an essential element in "assured access". Ironically, the US delegation report that followed the publication of the ICNT in 1977 neatly illustrated the essential interconnectedness of the whole system:

"in a sense, the negotiation is similar to playing a slot machine: all the cherries must be lined up in order to win" (p.19).

But dispute settlement was apparently not one of the cherries.

This may exaggerate the extent of Western acquiescence in the dispute settlement provisions of the ICNT. The US delegation report for the first part of the eighth session, in 1979, claimed (p.25) that, in the previous year, when negotiating groups were created to deal with "hard-core" issues,

> "it was generally understood... that the relevant texts, section 6 of Part XI and related annexes, were in need of revision",

though action had to wait for the eighth session itself when a "group of legal experts on the settlement of disputes relating to part XI", in form subordinate to the chairman of the First Committee, was set up under an East German chairman, Harry Wuensche. It proved remarkably successful.

Before the Geneva part of that session was over, the chairman was able to suggest, as "providing a good basis for further negotiations", several important amendments, including a more careful formulation of the limitation of the SBDC's jurisdiction (Article 191 - now Article 189) making it clear that, where a rule of the Authority, in a given case, conflicted with the convention itself or a contract made under it, the Tribunal would uphold the latter. Though Richardson would have liked further drafting changes, his delegation report on the session a year later commented that

> "it was generally agreed that <this article> means that the SBDC should not apply rules, regulations or procedures where doing so would violate the convention, and that it should not enforce allegedly discretionary acts of the Authority which constitute abuses or misuses of power"
> (US Delegation 1979, p.30).

Provisions for dispute settlement were strengthened in other ways. The right of a disappointed applicant for a contract to appeal to the SBDC, though covered indirectly in the ICNT (Article 187 2(b)), was made explicit, thus complementing the explicitness, noted in Chapter 7, of the formulation of the grounds on which such an application could be refused. Secondly, any State Party, or "natural or juridical person", having a contract with the Authority, which claimed that an employee of the latter had leaked 'secrets', data, or confidential information to its detriment, could insist on the Authority bringing its employee before "an appropriate tribunal" (not necessarily the SBDC) and the Party concerned (as contractor or sponsor) could participate in the proceedings (Article 168.3).

The mode of appointing the SBDC was also changed. The International Law of the Sea Tribunal (ITLOS) was itself given that task, and the role of the Assembly was limited to "recommendations of a general nature". (Annex VI, Article 35, p.2) It might still favour developing countries, but it would be somewhat more insulated from politics than it would have been under the ICNT. Moreover, for inter-state disputes, recourse could be had by either party to an _ad hoc_ chamber of the SBDC, constituted by the parties, with the SBDC chairman adding a deciding member if the parties were unable to agree.

More importantly, certain types of case were removed from the jurisdiction of the SBDC. Mention has already been made of the reference to an appropriate tribunal in Article 168.3, which may not be ITLOS or its SBDC. For disputes over the interpretation of a contract, it was agreed at the Spring 1980 session that either party could secure the referral of a dispute to "binding commercial arbitration", conducted in accordance with the UNCITRAL Arbitration Rules unless the parties agree, or the Authority in its rules prescribes, otherwise, (Article 188 2(a) & (c)). Similar arbitration had been agreed for disputes over the price at which a contractor might have to sell technology to the Authority under Article 5.3(e) of Annex III.

In the former event, however, the proviso was made that if the tribunal found at any point that a case involved a question of interpretation of the convention itself, it should refer that question to the SBDC, and decide the case in conformity with its reply.

Dispute settlement did not feature much in the Reagan review's criticisms of the convention. The "Approaches" paper (p.18) mentions "improving the composition of the Sea-bed Disputes Chamber" by making the selection of its members a matter for a Council vote, but only in passing, as one of several possible solutions to the "danger" of creeping extension of the Authority's discretionary powers. The Green Book proposed simply to remove the article (189 in L78) limiting the SBDC's jurisdiction with respect to the decisions of the Authority, thus cutting the Gordian knot discussed earlier, and (perhaps accidentally, since there is no immediately obvious rationale for it and no explicit indication of an amendment at this point) omit paragraph 2(c), which specifies UNCITRAL rules as normally applicable to commercial arbitration, from the previous article (188). The seven power amendments - i.e. those formally proposed by the USA and its six ocean mining allies in the final stage before the convention was adopted, included virtually nothing on the subject. Apart from a slight change in the options open in inter-state disputes, their sole proposal was, not the abolition of Article 189, but merely a clarification of its wording, so that it was clearer that it meant what Richardson, in 1980, had said everyone agreed that it had meant.

There was no mention of dispute settlement in the compromise proposals of the "Friends of the Conference" and the text, as adopted, was unchanged from 1980.

As it now appears in the convention, the dispute settlement mechanism for sea-bed dispute, though not perfect, is workable. The article originally proposed by Mexico (Article 189: Limitation on jurisdiction with regard to decisions of the Authority) now reads:

> "The Sea-Bed Disputes Chamber shall have no jurisdiction with regard to the exercise by the Authority of its discretionary powers in accordance with this Part I; in no case shall it substitute its discretion for that of the Authority. Without prejudice to article 191, in exercising its jurisdiction pursuant to article 187, the Sea-Bed Disputes Chamber shall not pronounce itself on the question of whether any rules, regulations and procedures of the Authority are in conformity with this Convention, nor declare invalid any such rules, regulations and procedures. Its jurisdiction in this regard shall be confined to deciding claims that the application of any rules, regulations and procedures of the Authority in individual cases would be in conflict with the contractual obligations of the parties to the dispute of their obligations under this Convention, claims concerning excess of jurisdiction or misuse of power, and to claims for damages to be paid or other remedy to be given to the party concerned for the failure of the other party to comply with its contractual obligations or its obligation under this Convention".

Though this sounds, at first glance, a severe restriction on the SBDC's scope, read with article 187(b), which expressly gives it jurisdiction over disputes between a State Party and the Authority concerning "acts or omissions of the Authority or of a State Party alleged to be in violation of this Part or the Annexes relating thereto", it seems to bear out Richardson's contention of 1980: in a case where the Authority's rules conflict with the text of the convention or the terms of a contract, the SBDC must decide in favour of the latter, though it was not its job to say whether or not there might be other cases in which the same rule might validly operate.

"Dispute settlement" as conceived here, which might otherwise be called judicial review, is important because through it a decision of a Sea-bed Authority would lose the legitimacy it would otherwise have for its members. States, and their companies, have the power and are likely to keep it; but they would look arbitrary if, having joined an organisation, they defy its decisions, even if they claim that it has exceeded its powers. All recalcitrant states can expect to claim likewise, and in the absence of an effective avenue of appeal, the merits of such claims are difficult to assess. With such an avenue, and provided that it conforms to the judgement of the body concerned, defiance loses its arbitrariness. Insofar as the Law of the Sea Tribunal proves to consist predominantly of developing country nationals (individuals, of course, not representatives), as is likely, its jurisprudence will also be likely to have a developing country tinge; but that is a long way from having no jurisprudence at all.

NOTES

1. Chapter 7 above.

2. Among them Switzerland, Germany and Greece. Iran also went into it at some length, advocating a flexible system in which a party complained of had a choice of forum, but had to choose one and accept its decisions as binding.

3. Note by the Secretariat to participating delegations, 28th April, 1977, p.39.

4. A/Conf. 62/WP.9/Rev.2, Art.15.

11 Pioneer Investment and the Interim Regime

The straightforward way of establishing a regime for the sea-bed, and a Sea-bed Authority to administer it, would have been to negotiate a convention, specify how many states, and perhaps of what category, would have to ratify it, to bring it into force, and then wait for these ratifications. A more sophisticated, and not uncommon, variant of this would be to provide for provisional application of the convention on signature and for the establishment of a body which would organise the organs and staff of the Authority in advance so that, as soon as the convention came legally into force it was in a practical position to embark on the tasks the convention had given it. In the case of the Law of the Sea Convention, however, it was agreed that even this was insufficient, in two main respects. In the first place, most mining states, without whose participation the Sea-bed Authority would exist only notionally, wanted to know, before they ratified, what the Authority's "Rules Regulations and Procedures" would look like. By ratifying they would be assenting, inter alia, to Article 137.3, which prohibited them from making any claims to the minerals of the Area "except in accordance with this Part", i.e. Part XI. Some, if not all, of them were unwilling to make any such commitments until the rules of the game had been agreed even more precisely than in the ninety-nine articles of Part XI of the convention and its related annexes. Thus for their point of view the most optimistic programme would have been, first to negotiate an acceptable convention, with a Preparatory Commission to prepare for its coming into force, secondly to sign the convention, and help to set up the Preparatory Commission, thirdly to negotiate with the commission a satisfactory set of "Rules, Regulations and Procedures" to be adopted by that body which would be difficult for the Authority subsequently to change, finally, to ratify the convention and so, when enough others did likewise, bring it into force. Thus the process of negotiating rules in the commission would be an essential element in determining whether the Sea-bed Authority became a reality. A resolution was needed, therefore, which, at the same time as the convention was adopted, established this Preparatory Commission and outlined its functions, composition and voting system. This resolution was one of the last items with which UNCLOS III concerned itself and we shall return to it later in this chapter.

By itself, though, this was still not enough. UNCLOS III had lasted so long that some leading consortia had already spent considerable sums on ocean mining investment, some of it site-specific. They had been encouraged in this by the passage of national legislation in the USA, and other countries, from 1980 onwards, purporting to grant their companies licences conferring, on nationally-specified conditions, exclusive rights to mine extensive tracts of the "common heritage". Even before the first legislation had been passed Richardson had warned the conference that the system of the convention, which asserted the exclusive jurisdiction of the Authority over all mining in the Area, would somehow have to be integrated with these nationally recognised claims, and had tabled an informal proposal to that effect. The convention as we have seen not only imposed an extensive list of conditions (1) on the award of a contract, but also required that an applicant must first let the Authority choose between two sites and be prepared to mine whichever one it did not choose, a requirement which could not be met if an investor had already started to mine a site. Subject to these requirements, and restrictions, applicants were assured of "access" and the right to mine, but their rights to a specific site began only when a contract had been signed. Investors convinced their governments that, if they were to have any confidence in the convention system, they needed an early international blessing on the preparatory work they were doing on a chosen site, a blessing that could in due course - when the convention entered into force - be converted into a contract to mine.

The outbreak of national legislation of 1980 onwards reinforced the credibility of the implicit threat of capitalist mining states that if they did not get the kind of mining system and the kind of recognition of site specific investment, that they wanted, they would stay outside the system. Though the legislation may have been ostensibly "interim", pending the entry into force, for the state concerned, of an international system of exploitation, these states would clearly suffer a loss of commercial reputation if they agreed to a system that gave no recognition to any licences they had issued.

Hence the need for what came to be called Preparatory Investment Protection (PIP). Though the Group of 77 repeatedly condemned US legislation, and the guarded support it received from Richardson after the ICNT fiasco in 1977, during its long passage through Congress, they did not oppose PIP. It was scheduled as one of the remaining items to be cleared up at the tenth session in 1981, but, in the face of the dislocation produced by the Reagan review, never discussed. The US then withdrew Richardson's informal paper of 1980. Essentially, negotiation of the question was confined to the eleventh session.

Negotiations began, behind the scenes, very early in the 1982 session. Ratiner (1982, p.1014) attributes this to the psychological subtlety of the Group of 77. They "sensed", he says, that there were those in the USA who did not <u>want</u> to see the convention improved, because they wanted the USA to stay out of it at all costs and to persuade its allies to join it in doing so. They therefore insisted on demanding

"an opportunity to negotiate the one issue on which they were prepared to make a concession so significant as to lure our allies into the treaty. ... Moreover, ... the developing countries hoped that if they made meaningful concessions on grandfather rights the US mining industry would be pacified and would reduce its pressure on the US government. In turn, they assumed the United States would reduce its demands."

Since the USA was at that stage precluded by the tightness of its instructions from making any concessions on any item before the conference, Ratiner argued, they were in no position to forestall this initiative or prevent the conference considering it at the outset of the session.

It was first raised officially in a note by President Koh of March 8 1982. (2)

Having sounded out some unspecified industrialised states and "participants representing the developing countries," he reported a "most constructive" exchange of views, and advised against scheduling debate on the question that might interfere with the "intense negotiation" needed to achieve agreement. He identified two issues that had emerged. One, a concern about the possibility that the Enterprise would require a monopoly of sites in the Area if each pioneer investor had to submit two sites to it, seems to have been confined to one state (presumably the USA) and to have been an attack on the whole system of exploitation under the convention; the other, of which more was to be heard, was over whether PIP could cover actual exploitation, or only the exploration of a site.

The note invited proposals and four followed: one from the Group of 77; one from a group of four: The Federal Republic of Germany, Japan, the UK, and the USA: one from Australia, Canada, Denmark and Norway on behalf of ten "Friends of the Conference" attempting, as on other questions, to reconcile these two groups: and one from France, the main feature of which was a set of suggestions for requiring pioneer investors to assist the Enterprise, by jointly training its staff and exploring for it the site it proposed to exploit first. France proposed that a pioneer investor should be able to exploit its own site only when it had done its assigned share of this for the Enterprise's first site.

The Group of 77's proposal contained several stipulations as to the granting of PIP. It would be offered only to signatories of the convention and their nationals. These states would therefore have to sign first and then apply for or sponsor applications for pioneer status, rather than vice versa. Like applicants for a contract in the convention itself, pioneer investors would have to offer the Authority, in the form of the Preparatory Commission, a choice of two areas, and as in the French plan make personnel training, capital and technology available to the Enterprise, and, on a cost reimbursable basis, help to explore its first site. Exploitation, as distinct from exploration, would have to wait for the convention coming into force and the grant of a contract under the sponsorship of state becoming a party to it; and while pioneer investors would have "priority" over other applicants in the grant of contracts to them by the Council, there would be "no automaticity" about such grants.

The USA and its allies, on the other hand, who wanted to make the resolution a protocol, having legal force independently of the convention itself, spoke of "certifying States" who did not need to be signatories of the convention though they needed to have ratified it within a year of its coming into force to remain "certifying States". They also made clear their intention to permit, in that case, a pioneer to seek sponsorship of a "state" that was a party. By this proposal, title to the areas as large as 150,000 sq km began to be spoken of (at Caracas, eight years earlier the size suggested in the American proposal had been only 30,000 sq km). The Enterprise was not to be given a choice between two sites, but each certifying state undertook to give it an area of "equivalent size and comparable value" to that it was itself exploring - an undertaking not very easily protected from abuse. Certifying states (not the investors themselves) accepted the (somewhat vague) obligations of states parties to the convention to facilitate the transfer of technology to the Enterprise. All "pioneer investors" would have had to do would have been to provide training opportunities for developing country personnel, and then in return for "adequate compensation" the Authority had to approve contracts for previously registered pioneer investors (a matter requiring amendment of the convention) whereupon the certifying state became a sponsoring state; the pioneer investors' obligation to accept the rules, regulations and procedures of the Authority was subject to their not imposing "significant new economic burdens" (note to para 3). Finally, the proposal made an important distinction between a "pioneer operator" which had spent at least $30m, and at least 10% of that on site-specific pioneer activities, before 1 July 1980, and "pioneer investors" which met these conditions only after that date, but before 1 January 1983; the former were to have priority in the allocation of these potentially enormous sites.

These were in many ways extreme proposals. The Group of Ten's "mediating" proposal had two similarities to them, one more apparent than real. They too envisaged two categories of pioneers, but the priority would be enjoyed by all who had spent the necessary sums by 1st January 1983, which for "the US and others" was a requirement for all pioneers. The Group of Ten's second category was that of "interim explorers", not to be considered before 1st January 1985, which covered a state or state-sponsored entity certified to be embarking on such investment; the status would lapse unless, within six years of registration, the sponsoring state certified that it had met the condition. The other, genuine, similarity with the mining states' proposal was that they explicitly allowed for "flags of convenience" - pioneers changing their sponsors to keep within the Authority's system. In several other respects they inclined towards the Group of 77 position: sponsoring states had to be signatories; pioneers, like contractors in the system proper, would have had to offer the Prep Com a choice of two sites, and would also have had to train personnel for the Enterprise free. They would not, as pioneers, have been permitted exploitation. That would have had to wait for the Authority's grant of a contract when the convention entered into force, in which they would have enjoyed priority, but no automatic entitlement.

On March 29 Koh produced his own text, (3) introducing it with an explanation which was a model of lucidity. He accommodated American demands to the extent of allowing for pioneer areas of up to 150,000 sq kms; committing the Authority to approve a contract with all pioneer investors; and closing the door to pioneer investors by January 1st 1983, though there was to be no hierarchy among them.

He followed the French idea of requiring the pioneer to help the Enterprise with training, and with the exploration, on a cost reimbursable basis, of the latter's own first site. Additional obligations, in line with Group of 77 proposals, were imposed on investors to supply the Authority with technology and "the necessary funds". The first of these obligations corresponded to a familiar, if highly contentious, part of the convention's system of exploitation. The second corresponded to nothing in the text and read oddly. Five of his other proposals met Group of 77 demands: first, the certifying State must be a signatory of the convention; secondly it must offer the Enterprise the choice of two sites; thirdly each investor would be entitled to one area only; fourthly, the convention's production ceiling would govern production authorisations to which the pioneers would become entitled when they duly obtained contracts from the Authority; and fifthly, the rights would lapse if the convention had not entered into force five years after its adoption. (The first two of these also reflected, of course, the views of the Group of Ten).

When, on April 1st, Ambassador Malone responded to these proposals for the USA, he recognised the concessions that had been made to ocean miners' demands but listed seven specific deficiencies in them. First they violated "a consensus in informal negotiations" to distinguish between early and late investors; secondly they mentioned "production limitation" (a notion then anathema to the Americans); thirdly by cutting off pioneer rights after five years if the convention had not come into force they denied investors the necessary security; fourthly they did not guarantee confidentiality of data handed over to the Prep Com, fifthly the obligations on pioneers to assist the Enterprise were too burdensome and should be transferred to their certifying States; sixth the registration fee ($500,000) and performance requirements ($1m) were respectively "improper" and "unrealistic"; and finally they did not meet the need to exempt pioneers from any rules, regulations and procedures whereby, by imposing significant new economic burdens on them, would render continuing operations uneconomic (Malone 1982).

The third of these objections was immediately met. The next day Koh issued a revised version whereby the protection the resolution offered would remain until the convention entered into force. Soon after, on April 7th, it was opened for formal amendment, together with the other resolutions and the convention itself. Of thirty one sets of amendments submitted (A/Conf. 62/L96 - 126), six concerned PIP:(L97 - Gabon; L105 - Japan; L106 - France; L116 - Peru for the Group of 77; L122 - Belgium, Germany, Italy, the UK and the USA; and L125 - the USSR). This pattern of sponsorship suggested that Japan, like France earlier, was beginning to detach herself from the "hard core" of ocean mining states, and that Belgium and Italy had joined it - though all seven, as we have seen, co-sponsored a wide ranging series of amendments to the convention itself. The fact that, of the seven, France and Japan (only) went on to vote for adoption of the convention, and sign it in December, may offer some indication of how important PIP was.

The hardcore group did at last accept that the principle, enshrined since 1975 in the Basic Conditions of Exploitation, of requiring applicants to a contract to offer the Authority a choice of two sites, should apply, mutatis mutandis, to PIP.

They also accepted the applicability of the convention's production limitation formula, and, tacitly, the application of the Authority's "rules, regulations and procedures" to pioneers, equally with later applicants, once the convention was in force; but America's four co-sponsors sought to give effect to Malone's remaining five objections (the first, third, fourth, fifth and sixth) and, in the case of the third, which Koh had already changed the text to accommodate, to go beyond them. Delay in bringing the convention into force would represent continued disagreement. Koh, attributing such disagreement to mining states, had originally suggested that its consequences should damage only the investors (by terminating their PIP rights); his revised text of April 2nd had moved to a neutral position. Pioneers would continue to enjoy that status, but could not begin exploitation. The five now wanted to entitle them to begin commercial production if the convention had not entered into force by 1 January 1988, thus giving them no incentive to bring pressure on their sponsors to ratify it (although, legally, the convention would enter into force after the ratification of _any_ sixty states, in practice these would probably be forthcoming only if it was also widely ratified by ocean mining states).

There was general agreement that some opportunity should be given for developing countries, or entities therefrom, to gain pioneer status. The five did not agree to the Koh formulation, but allowed for the expenditure requirement to be further negotiated. (This was perhaps a demonstration of, even possibly a conscious protest against, what they saw as the premature curtailment of negotiations. Though technically amendments which would, if accepted, supposedly have produced what they regarded as a better PIP Resolution, they would actually have left it self-evidently incomplete). Japan's amendments included one which would have required the same amount of investment, in real terms, from developing country pioneers, but have left the door open to them until 1 January 1985. The Group of 77 asked for a similarly extended deadline for their own pioneers (as well as lower investment requirements). Gabon wanted to impose on developing countries the same investment requirements as applied to other states, but to remove the stipulation that a part of it be site-specific. France, by contrast, wanted to put the qualifying date in the past, 1 January 1982, thus closing the door completely to new pioneers, developing or otherwise.

Japan and the Group óf 77 agreed on one further point: Koh's "pioneer areas" were too large. Both proposed that the upper limit be 60,000 sq km (instead of 150,000); alternatively, the 77 would have left it to the Prep Com to fix. They would also have made some other important changes to PIP. One would have been to equip the Prep Com with the power to verify the expenditures pioneers and their States claimed to have made; another to require them to invest $10m a year, and pay the Prep Com $1m payable on obtaining a contract. Further, an investor's convenient change of nationality would have been made conditional on the newly-sponsoring state's having effective control over him; he would be liable, with respect to transfer of technology, to fulfil the obligation of a contractor under the convention; Finally (and in contradiction to a provision of the amendments submitted by Gabon), the resolution was to apply only to nodules, all other resources being governed by the convention. The USSR's only amendment to PIP (though not the only aspect of it about which it expressed disquiet) was to reaffirm the application to it of the convention's anti-monopoly clause.

On April 22, nine days after all these amendments were tabled, President Koh published a new set of proposals. (4) Their most striking feature was that instead of defining pioneer investors, the new proposals named eight of them: four consortia, and four state enterprises (those of France, India, Japan and the USSR). Only developing countries could join this exclusive club, and they had until 1 January 1985 to qualify.

The size of the pioneer area was kept at 150,000 sq km. Since every applicant had to offer the Authority two such areas, this meant that the eight pioneers, even if there were no more of them, could dispose of 2,400,000 sq km between them. Pioneer investors were, however, required to relinquish in three stages extending over eight years, at least half of the area assigned them, and more if, as now seems unlikely, the Authority, in its Rules, Regulations and Procedures, prescribes a maximum size for sites smaller than 75,000 sq kms. Disregarding Malone's strictures on the point, Koh incorporated the 77's proposal for charging each pioneer an annual fixed fee of $1m, but made it payable on eventual approval of his plan of work by the Authority, for which the pioneer would need to apply within six months of the convention's coming into force. Only signatories of the convention could sponsor applicants for pioneer status, and before a pioneer could get a plan of work all the components must have the nationality of States Parties; but investors (i.e. consortia as a whole) could change their nationality and sponsorship if their original state did not qualify, provided that the state to which it transfers has effective control over it. Elaborate procedures were laid down for ensuring that pioneers agree on the boundaries of their site, some priority, as the Americans demanded, being given to early investors, and that they are assured, under the Authority, of production authorisations, subject always to the production limit. As the growth segment of the nickel market, assuming there still is one, increases, investors ready to begin commercial production first would have first claim to authorisations, and after them, the Enterprise.

Koh accepted the American concern about confidentiality of data submitted to Prep Com and the Authority and included an amendment to that effect; and further lightened the burden on investors by transferring, from them to their certifying States, the duty of assuring the Enterprise of "the necessary funds" for keeping pace with them; but his new text still left them obliged to train the Enterprise's personnel, conform to the convention's technology transfer obligations, and, on a costs-plus-interest basis, assist the Authority in exploring its first site. Koh could hardly offer the Enterprise less than France had proposed to.

This new version of PIP clearly did not satisfy all the mining countries, though it did some. It also encountered opposition from the Group of 77, because of the size of pioneer areas, and from the USSR on two grounds, one being that it was improper for the conference to name companies and give them, as such the status of pioneer investors. (It was presumably less improper to name states). The USSR further objected to the survival, albeit in stricter form, of the option of pioneer investors to change states. Thus, it claimed, while its own pioneer status would depend on its becoming a signatory of the convention, the activities of American firms would enjoy that status without America incurring the obligations of signatories, by a change in the nationality of the consortia to which they belonged.

Koh's answer to this was that the industrialised states had made "an even greater concession": that a consortium's plan of work could not be approved by the Authority unless all its components were States Parties or nationals thereof. This is indeed an important concession. It implies for instance, that if Belgium and Italy ratify, and the USA does not, and the Convention comes into force, Ocean Mining Associates could obtain a contract only if "Essex Minerals", a subsidiary of US Steel, took out Belgian or Italian nationality, (or dropped out of the consortium). Nevertheless, Koh's answer did not satisfy the USSR, and on that account, with her allies, she abstained on the adoption votes.

Otherwise, the main pressure for change in PIP came from the Group of 77; particularly for the reduction of the maximum size of sites. To avert a vote on this, Koh persuaded the industrialised states to agree to reserve to the Enterprise (after the pioneers had had their chance) production authorisation for two sites; and on April 29 he introduced a further amendment to this effect.

The resolution was adopted in its amended form. Its effect was to offer pioneers a remarkably attractive deal, (if only their governments signed, and went on to ratify, the convention), and, at the instance of ocean mining states, to supersede, for the benefit of a privileged few, the principle of freedom of access for which those same states had ostensibly contended so strenuously for so long.

As Ratiner, on this point, concluded

"The negotiations ... resulted in a final Conference Resolution which successfully met some of our fundamental objectives - but more importantly may well have met the most central objective held by our closest allies ... altogether, ten sea-bed mining entities are entitled to all of the mineral production likely or possible from the sea-bed for the next 30 to 50 years.

Thus - with the notable exceptions of mandatory technology transfer and the procedure for amending the treaty - the offensive ideological provisions of the treaty would not effectively apply before the middle of the twenty-first century" (Ratiner 1982, pp.1014-1015).

Apart from the named eight, plus any developing country qualifiers by 1 January 1985, investors in sea-bed mining would have to wait until the convention was in force, and the pick of the sites had been snapped up. Of what now seems the most promising 2,400,000 sq. km, or more, of the Area, half would, under the PIP resolution, go to the Enterprise as "reserved", and presumably only the less attractive portions of the remainder would become, through relinquishment, open to applicants. Late-comers, even if they could find a site that looked likely to prove profitable, would have to queue for production authorisations behind the pioneers and the two "Enterprise" sites. If any group of states had cause for complaint, it would be the non-pioneer industrialised states who would have little chance of participating in ocean mining, who would have to wait inordinately long for a seat on the Council, and who were not likely to be substantial beneficiaries from revenue sharing, yet would have to pay and guarantee, in proportion to their share of the UN budget, the initial loan to the Enterprise. Yet only two members of the Western European and Others Group who were not named among the pioneers (Luxembourg and Spain) abstained on the vote; and only Israel, for quite different reasons, voted against.

The Preparatory Commission

As mentioned earlier, the Preparatory Commission (Prep Com) was an integral part of the negotiation process, because the ocean mining states had made it clear that their firm commitment to a comprehensive treaty system would be conditional on what emerged from Prep Com. It was agreed that Prep Com was to be established by a resolution of the conference, passed at the same time as the convention was adopted. There were arguments about its composition, functions, decision making system, financing, and the timing and mode of its transfer to the Sea-bed Authority itself; but there was little ferocity in them, since unlike PIP and the convention itself, Prep Com represented, in essence, a continued opportunity to negotiate, rather than the outcome of critical decisions, and it was generally realised that there was little point in giving (or denying) Prep Com the capacity to make majority decisions, since for the convention to achieve universal or near universal participation all major groups, and particularly the ocean mining states and the Group of 77, had to be reasonably satisfied with Prep Com's results.

Discussion had been initiated in Amerasinghe's presidency in March 1980, and was one of the few items of business with which the tenth session, in 1981, concerned itself, while waiting for the results of the Reagan review. A further paper by Engo in the August of that year formed the basis of the proposals jointly tabled by Koh and Engo at the eleventh session on 19 March 1982 (A/Conf. 62/C.1/L30). Membership of Prep Com was to be confined to signatories of the convention, but all who signed the Final Act could participate as observers without voting rights. It was to be financed from the United Nations budget which meant all UN members would be asked to pay for it whether or not they were members. This attracted some opposition from those uncertain whether they were going to be sufficiently satisfied with the convention that had still to emerge in its final form (and the USA went on to refuse to pay its share), but this can hardly been crucial to decisions on signature. For decision-making, the initial arrangements of Prep Com were to follow UNCLOS III's own Rules of Procedure. Through those rules, it could adopt its own decision-making system. The functions of Prep Com were to prepare the agendas of the first meeting of the Authority's Assembly and Council, to adopt "rules, regulations and procedures" to govern mining in the Area, to make the necessary preparations for the Enterprise to be able to start up promptly in business, on entry into force of the convention, and assume the responsibilities already mentioned in connection with PIP, of administering the "private and state" track of the parallel system during the period between signature of the treaty and the establishment of the Authority.

One important aspect of the Prep Com's role was not defined in watertight terms. The rules would be draft rules and it would be open to the Council (by consensus as the treaty prescribes) to modify them. It was generally understood, though not made explicit in the resolution, that rules drawn up by the Prep Com would be adopted by the Council in the absence of a decision to the contrary. Thus a state satisfied with the Prep Com's rules would have some assurance that the Authority would proceed without further delay on this basis. Although this is not indicated in the resolution, it would be open to Prep Com to suggest a protocol or amendment to the convention, embodying agreed rules. Thus if the rules had once been agreed, states could know, at the point of joining, and thus of committing themselves legally to the convention, that the Authority was already itself legally committed to these rules.

The resolution also addressed itself to how and when the Prep Com would terminate. It would report to the first session of the Assembly and terminate at the end of it, but it was made clear that this did not give the Assembly, in acting on the Prep Com's report, any power to exceed the jurisdiction conferred on it by the convention. Finally, the Prep Com was also given the task of preparing for the establishment of the Law of The Sea Tribunal, for submission to a meeting of states parties convened for this purpose. It was to meet at the seat of the Authority, Jamaica.

Several aspects of these proposals for Prep Com came under fire, together with the other resolutions produced with it in L30 and the convention itself, from the US Delegation leaders, Malone, in his speech of April 1 1982. He wanted full membership to be open to signatories of the Final Act and for decisions to be taken by an Executive Committee constituted in the same way as the Council and by the decision making procedures set out in the Council. Since, however, these criticisms were accompanied by such a discouraging assessment of the convention itself (and PIP) from the US point of view, so that the chances of US participation in an eventual treaty were seen as more dependent on these other factors, the objections did not carry much weight. Koh made no further changes in his proposals for Prep Com, and the USA went on to say that not only would it not sign the Convention, it would not take up the observer status conferred on it by the signature of the conference's Final Act. When in March 1983, the Preparatory Commission met for the first time, the USA, which had done so much to shape the regime and machinery for the common heritage of mankind, was represented, as on an occasion sixty years earlier of which something similar could be said, by an empty chair.

NOTES

1. See Chapter 7 above.

2. IPIC/1 "Treatment to be accorded to preparatory investments compatible with the convention". The issue had initially been brought before UNCLOS III in the informal working paper of the U.S., Document 1A/1 of April 2nd 1980, and then described in A/Conf. 62/BUR.13/Rev.1.

3. A/Conf. 62/C.1/L30.

4. A/Conf. 62/L132/Add.1.

CONCLUSION

12 Outcome and Prospects

A finished history of the negotiation of a regime for the sea-bed is almost as distant a prospect as a finished history of the world. What UNCLOS III left undone when, with the adoption and signature of a convention in 1982, it came to an end, the Preparatory Commission will seek to do; and it may even try to undo, or amend, some of what its predecessors had done. This moment of transition may nevertheless be a good point to pause, and survey the fruits so far, in the light of the hope and expectations with which this whole undertaking was launched.

It must be said at once that the harvest has been long in coming, far longer, already than anyone originally envisaged. When Dr. Pardo, in 1967 warned his General Assembly audience that "(our long-term objectives) cannot be achieved either quickly or easily" (Pardo 1975, p.41) he contemplated (p.39) having to wait for their achievement only to 1970; By 1970, the General Assembly looked to the Third Law of the Sea Conference as being convened no later than 1973. By 1973, it was clear that the session held that year would be confined to procedural questions, and that UNCLOS III might well need two substantive sessions, in 1974 and 1975, to complete its work. At the first of these, at Caracas, this was widely seen as the most generous schedule conceivable. The leader of the American delegation, John Stevenson, addressing the session on July 11th was typical of many when he said

> "It is the view of my delegation that the Conference should strive to adopt an entire treaty text this summer. What is required ... is ... the political will to decide a relatively small number of critical issues ... If we do not at least try to reach agreement on the treaty this summer, we may well not even achieve the basic minimum required to finish next year and in the interim prevent further unilateral action prejudicial to the success of the Conference" (Stevenson, statement to plenary 1974).

And, though, after the third session, there was a fourth, and so on until the eleventh in 1982, the next one was always spoken of as the last. Though we may conclude, with hindsight, that the complexity of the task of negotiating a sea-bed authority made it impossible for UNCLOS III to complete it much more quickly than it did, that was not the common view at the time.

It has also long been obvious that the domain a Sea-bed Authority would eventually inherit - the International Area of the sea-bed - will be much smaller than was generally assumed in 1967. Pardo (1975, p.39) imagined that coastal state jurisdiction would be defined "approximately at the 200 metre isobath or at twelve miles from the nearest coast". Instead, the "continental shelf", has been defined so as to include the 200-mile exclusive economic zone and extend far beyond the foot of the continental slope. It is true that, where as even in 1967, Pardo (1975 p.22) was able to cite Shigeru Oda's interpretation of the 1958 convention as implying that "all the submerged lands of the world are necessarily parts of the continental shelf by [its] very definition", the 1982 convention set limits to the shelf and established an international body to monitor coastal state claims; yet not everyone, in 1967, agreed with Oda's reading of the legal position; and at the very least one can say that the prospect of making the international area of the sea-bed the "common heritage of mankind" did not induce the world's coastal states to retract their claims to jurisdiction in its favour, as Pardo himself has supposed it might. Rather, the negotiating process to which Pardo's proposal gave rise seemed to add urgency to the process of "enclosure".

The most striking feature of the system of exploitation that a sea-bed authority will administer is its elaborateness. The West wanted assured access for its private and often multinational consortia; the USSR wanted the same for the public enterprises of socialist states; and the developing countries wanted to endow the Authority with managerial and entrepreneurial powers. The resulting compromise has taken the form of a parallel system. In return for equipping the Authority, through the Enterprise, with the capacity and the finance to set up in the business of ocean mining, the mining states, and particularly those of the West, have insisted that their own (private or state) miners should be equally assured that they will be able to mine the sea-bed on "reasonable" terms, and that the convention must thus embody, in precise form, the grounds on which the Authority can refuse a contract (or a production authorisation) and the conditions, including the financial obligations, it can impose on contractors. It can be expected that the already voluminous provisions of the convention which thus resulted will be considerably augmented by the rules, regulations and procedures which the Preparatory Commission is now engaged in negotiating - a monument to the distrust felt among prospective mining states - the West, in the end, even more than the USSR - of a Sea-Bed Authority endowed with significant discretion. The elaboration of this regime has been amply seasoned with "interimitis". (1) There is to be an "interim period", in which sea-bed production will be subject to a production ceiling, and at the close of which, after a review conference, the system may be amended, though without affecting what we listed as fundamental principles; by the votes and ratifications of three quarters of the states parties. Moreover, even before all this, there is a system for the interim protection of pioneer investors, by which the Preparatory Commission, having designated one of two areas submitted to it by a pioneer, as reserved for the Enterprise, authorises the said pioneer to explore the other with an assurance that he will get a contract to exploit the better part of it when the convention comes into force. These areas are huge - 150,000 sq km each. The pioneers named in the resolution concerned (which include one developing country India) will between them account for 2,400,000 sq kms, half of which will be retained by the Authority, and a further quarter of which must be relinquished within eight years.

Later would-be miners have to look elsewhere, or in the relinquished parts of pioneer areas, for their sites, and within the interim period, queue behind pioneers for production authorisations. Ultimately the "private" tract represents "freedom of access", the embodiment, in an appropriately modified form, of the vaunted principle of freedom of the seas. In the interim, it sanctifies the privileges of a few pioneers.

Finally, and sadly, UNCLOS III has not achieved the unanimity for which it strove. Of the ocean mining states, the USA voted against adopting the convention and has subsequently turned its back on it, refusing even to attend the Preparatory Commission. Belgium, Germany, Italy and the UK abstained on the vote and have not signed the convention, but have taken up observer status in Prep Com. The day when we have a generally accepted regime for the " common heritage" does not now seem imminent.

Why was unanimity not achieved? What would have had to be different for UNCLOS III to have arrived at a generally acceptable convention? Answers to such questions, all necessarily speculative, can be given at a number of different levels.

It is not out of the question that agreement could have been reached at the final session. The seemingly enormous chasm between the USA and the Group of 77 at the session's beginning had been reduced, by its close, to differences over a couple of points. There was some ground for the contention of some Western states that negotiations were terminated before all prospects of reaching consensus had in fact been exhausted. It could well be that had the US Delegation not been required by its instructions to seem to be insisting on wholesale changes and unable to make concessions until half way through the session; or had the Group of 77 found the flexibility to accept, at this late stage, important but not devastating changes in the text as a price of securing universality; or had the conference's time-table, for once, rigidly adhered to, been modified to permit some further time for negotiations, UNCLOS III would have ended, in 1982, on a triumphantly consensual note.

The Group of 77 did not, in 1982, use its numerical strength to make the existing text more to its liking. On the contrary, for the sake of promoting consensus, it made concessions which were reflected in the convention as adopted. The fact that the latter was signed by Australia, Canada, China, France, Japan, The Netherlands, the Scandinavian Countries, and the USSR shows that its attractions were not confined to developing states, which was not surprising since it was (from the Western viewpoint) a somewhat improved version of a text on which consensus had largely been reached in 1980. The sea-bed mining provisions of that text (ICNT Rev 3, alias the draft convention) was the product of a prolonged process of negotiation between, essentially, those who had the capacity to mine and those who had the votes. The course of the courtship did not always run smooth. There were serious set backs in 1975 (the SNT), 1976 (the RSNT) and 1977 (the ICNT). But between 1977 and 1980 the ICNT was sufficiently improved for the chief critic, the US Delegation leader, Elliot Richardson, to sound almost euphoric about the chances of adopting a convention, acceptable to his country, in the following year, to a descant of similar, if more grudging, expressions of satisfaction by other mining states.

Thus at this level the cause of failure to reach consensus was that what (largely) satisfied Richardson in 1980 did not satisfy the United States, at least once Reagan was elected, and perhaps even earlier if, as is sometimes claimed, the convention as it stood, if presented to the Senate, would have been defeated.

The difficulties posed for multilateral negotiations of long term questions by the potential discontinuity of US policy, and the apparent inability of UNCLOS III to take such possibilities adequately into account, have been examined in Chapter 5.

UNCLOS III was unfortunate in having to encounter one of the more dramatic reversals of American political history, which has reverberated throughout world politics ever since, and particularly in those global fora sensitive to Third World concerns.

The odds were stacked against adjustment to such a tremor. This raised the question, also considered in chapter 5, of whether UNCLOS III could have succeeded by concluding its work more promptly while America's more internationalist mood prevailed.

To some extent the explanation lies in the realm of political psychology. The "common heritage of mankind" was an idea of very wide appeal, but those it captivated, as so often happens, were unwilling to take their dedication to it to the point of giving up some freedom of action. Like Millament who "may by degrees dwindle into a wife", such a concept may dwindle into a practical proposition - but only "by degrees". The process cannot, beyond a certain point be hurried.

It cannot however be said that in UNCLOS III the process of transforming an idea into what was, up to 1980, a fairly acceptable compromise, reached its maximum speed. Institutional factors, in the Group of 77, in the USA even before 1980, and in the conference itself all contributed to delaying it.

The Group of 77 took time to organise itself in such a way as to negotiate as a group; yet this was an issue on which the group attached so much importance to its own solidarity as to preclude the adoption of any solution which did not have its support as a group.

It had difficulty in taking a position in 1974, and repudiated the RSNT, which its leaders had negotiated in 1976. By 1980, however, it was negotiating effectively:

"An important feature of the UNCLOS negotiations was that developing country representatives in the Group of 21 reported daily to the Group of 77 and the latter did not reverse any of the compromises or commitments it had made" (Commonwealth Secretariat 1982 p.58).

The US, too, spoke with more than one voice, even before 1980. Legislation on deep sea mining, fisheries, and pollution control seemed at times to cut across the delegation's negotiating position. Though there was little change in policy between the Carter Administration and its Republican predecessors, there was a widespread assumption, in 1976, that the elections would produce a more accommodating stance; and there was little appreciation of the hazards and delays that might separate a US vote for adoption of a convention and binding American commitment to it through ratification.

The procedures of UNCLOS III itself, and their probable effects, have been discussed extensively in Chapter 5. It is not that it failed to innovate; rather, the implications of some of its innovations were sometimes not fully anticipated. For instance, because the successive formulations emanated from committee chairmen (directly at first, and later through the collegium), the selection of such chairmen was crucial. Paul Engo, the First Committee's chairman, had energy and presence, but his judgement was in question. Twice (in 1975 and 1977) progress made under the

240

chairmanship of others (Pinto and Evensen respectively) was
negated by the alterations he took it upon himself to impose
on their results.

It has been said too, and by some not unsympathetic to
developing countries (e.g. Kirthisingha 1981) that, the
Group of 77, at UNCLOS III, became too attached to the
creation of a working Enterprise, and to the obligation of
contractors to transfer technology to it. The latter proved
a particularly difficult issue for the West, yet it seemed
unduly diffident to the Group of 77 not to believe that, if
the Enterprise were given enough capital, it could devise or
acquire effective technology of its own, and make it
available to developing countries. The Enterprise itself is
a remarkable concept - an international public mining
corporation. That the 77 were willing to fight so hard to
get it the resources it needed to go into business implies
an identification with common achievements unparalleled in
any other sphere of the Group's activity. Without it, the
convention would have been considerably easier to negotiate.

In spite of not being unanimously approved, the convention
was signed by 118 states at the conference's closing
ceremony at Montego Bay, Jamaica, on December 10 1982, more
than twice the fifty required to launch the Preparatory
Commission, which duly met at Kingston on March 15 of the
following year. By the end of that month it had received a
further four signatures and two of the sixty ratifications
required to bring it into force. Many who signed at Montego
Bay - including developing countries and land-based
producers - stressed the achievement of universality for the
convention as the Preparatory Commission's primary aim.
France, a signatory, emphasised the need to correct "the
imperfections with regard to technology transfer and the
financing of the Authority". (2) Clearly, if such attitudes
prevail, UNCLOS III will continue under another name; but
as things stand, its negotiations will be conducted in the
absence of the USA which, though it signed the Final Act,
and is thus entitled to participate as an observer, made
clear that it would not do so.

The convention offers one path to the legislation of
mining in the international area. Ocean mining states have
for some time contended that there is an alternative path
through national legislation. "Interim" legislation was
first passed by the US Congress on June 28 1980. By October
1982, five other countries (The Federal Republic of Germany,
The UK, France, The USSR and Japan) had done the same and
Italy was expected to follow shortly. (3) Four of these
(France, Germany, the UK and the USA) signed a "Conflict
Resolution" agreement on September 2nd 1982. In order not
to preclude operating within the law of the convention, it
provided not for mutual recognition of mine sites but only
for mutual consultations over claims that might overlap.

There is thus, at present, a duality about the law governing ocean mining. Title to mine sites may be granted outside the convention, by states, or inside the convention by the Preparatory Commission, and eventually, the Sea-bed Authority. Mining under the Authority will cost a company more in tax and other obligations, and probably subject it to tighter regulations (although, environmentally, the US licensing regime may prove it be quite strict). But national title to a tract of the international sea-bed, allegedly claimed in virtue of the freedom of the seas, is not, by itself, adequate, since any other state could by virtue of the same principle make a claim to all or part of that same tract. Agreement among a few states to avoid such overlapping claims is better than nothing but offers only a partial respite. Other states need take no notice of it. Of course if all likely mining states succeed in avoiding overlap, they may feel their title is reasonably secure, even though the convention and most of the other states in world do not recognise them. But things move fast in this field. India, one of the few recognised pioneer states (as opposed to consortia) was not thought of as a likely mining state a year or two earlier.

The uncertainty of title granted by national legislation would be all the greater if America were the only mining state to stay outside the convention. The conference President, Tommy Koh, has threatened to challenge the legality of any mining outside the convention both in the General Assembly and in the International Court of Justice. The latter's judgement is legally binding only in "contentious" cases, which it can hear only if the parties have given it jurisdiction. America's acceptance of the compulsory jurisdiction of the Court is still nullified, in effect, by the six word Connolly Amendment, which excludes cases falling within the domestic jurisdiction of the United States as determined by the United States. The General Assembly could however request an Advisory Opinion from the Court. Asked whether the Sea-bed Authority or the Prep Com was obliged to recognise sites granted without reference to it, the Court would almost certainly, and perhaps even unanimously, give a negative answer.

On the other hand, though non recognition would mean that the Sea-bed Authority was perfectly entitled to grant title to all or part of the same site to another applicant, it is not likely that some state, or private entity, would apply for a site overlapping with one claimed under US Law, and add the prospect of conflict with the US Government to all the other costs and burdens attending ocean mining, particularly under an international regime, unless it had actually discovered some rich deposit, within the area concerned, so that area had already been demonstrated to be distinctly better endowed, as a whole, than any as yet unassigned site.

Other sanctions against "extra Sea-bed Authority" mining, though conceivable, are even less likely. Nevertheless, so long as there is more than one regime, conflict, of some kind or other, is possible and becomes somewhat more likely as the size of site increases. Title is thus, in international terms, more than a little uncertain.

Achieving unanimity through the Preparatory Commission is not likely to prove easy. In theory it could prepare amendments, preferably by consensus, for adoption, at their inception, by the Council and the Assembly, but this would be an exceedingly delicate process. If the Group of 77 is prepared to see the texts it negotiated in 1982 modified at all, to meet the requirements of the mining states, it will be only if such modifications secure American participation, but the refusal of the USA even to send an observer to the commission deprives the latter of any indication, let alone an assurance, that any given set of amendments would be enough to do this.

A more promising path is through the commission's performance of the task officially entrusted to it, the drafting of the Authority's "Rules, Regulations and Procedures". Some of the changes sought by the mining states, and the Group of 11, in their formal amendments of 1982, would have clarified, rather than changed, the mode of operation of the Authority envisaged in the convention. There would be nothing inconsistent with the convention in making explicit, through the rules, that since the Authority's budget needs the support of both the Assembly and the Council, if the Assembly amends a Council recommendation it goes back to the Council (and vice versa); that the Legal and Technical Commission must decide on its recommendation with respect to an application to mine within, say, 120 days of the application's submission; that there is also a time limit on the Authority's selection of one of the two sites proposed by an applicant; that the Authority's criteria for qualification standards of miners, for according priority where necessary between competing applications for production authorisations, are objective and reasonable; that the Authority's procedures in dealing with prospecting exploring, and exploiting the sea-bed will be simple and expeditious; or that, over a specified period, the Enterprise will repay its debts. Even the provision requiring applicants to sell their technology, if necessary, to the Enterprise could be nullified if the Enterprise, through a freely-negotiated joint venture, were able to acquire equally effective technology independently, since it arises only if the Enterprise is unable to do this (Article 5 3(a) of Annex III).

What such rules regulations and procedures could not deal with would be the possibility of amending the convention, in the last resort by a three quarters majority, at the end of the review conference. Since, at any time, amendments can be adopted by the Council and the Assembly, the whole point of a review conference is that at the end of it, decisions can be taken that might not, at some other point be possible. The idea of such a review (or reviews) goes back to a remark of the then US Foreign Secretary, Henry Kissinger, in 1976. From then on the conference has wrestled with the question of what happens if the review conference does not lead to universal agreement on changes. The answer given by the convention is less revolutionary than that found in the constitutional instruments of some international bodies (The Constitution of the World Health Organisation, for instance, can be amended by a two thirds majority), and it expressly protects rights enjoyed under existing contracts.

The process of negotiating an international sea-bed authority constitutes both a unique unrepeatable moment in world history, and an example of a phenomenon, that in recent years has become, and is likely to remain, increasingly common - global conference diplomacy. In this latter guise, it has some important lessons to teach us.

This has been a process animated by an idea - that of the "common heritage of mankind", with its corollary of global management of global resources. Few have been willing to dispute the validity of the idea as such, and its general acceptance - most strikingly visible at the first substantive session of UNCLOS III at Caracas in 1974 - has aroused high expectations. At best, these expectations have been only partially fulfilled. The sea-bed's resources now seem less plentiful and more costly to extract; the fraction of them that remains in the international domain is smaller. The scope of a sea-bed authority is restricted, its legitimacy is not universally recognised. As a vehicle for the global redistribution of wealth, it may be scarcely more than a token.

The tendency of the United Nations (and other international organisations) to generate ideas that prove capable of being realised, if at all, only on a drastically reduced scale, can also be seen, for example, UNCTAD's plan for an Integrated Programme for Commodities, financed by a Common Fund. When first mooted, at the fourth UNCTAD in Nairobi in 1976, it was seen as entailing International Commodity Agreements for eighteen commodities supported by a Fund of $6 billion. In 1980, agreement, yet to be implemented, was reached on a Fund of $470 million. By 1982 there had been only one new agreement (for natural rubber) and four existing agreements (for cocoa, coffee, sugar and tin) had been renegotiated. In a more political realm, the idea of collective security embraced by the League of Nations (rather more unequivocally than in the United

Nations) committed members to preserve each others' territorial integrity and existing political independence against external aggression (Article 10), and to the total isolation, at least, of any state resorting illegally to war (articles 16 & 17). In practice, when Italy attacked Ethiopia in 1935, partial (and ineffective) sanctions were imposed on her by most League members.

The lesson of this pattern is that in evaluating a proposal for international action, it is necessary to ask whether it would still be worth supporting in the attenuated form in which it is likely to be implemented. What is the minimum scale of operations below which its realisation of a scheme is worse than useless?

Opponents of a sea-bed authority will claim that we have already reached the bleak predicament where potential miners will be deterred from mining by the burdens such an authority is likely to impose on them, so that either there will be no mining at all to generate revenues from the Authority, or that any revenues it receives will be swallowed up by the costs of its vast bureaucracy. It is not easy to discern the truth in either case, but the signs are that they exaggerate in both these respects.

All the states named as pioneer investors in their own right (France, India, Japan and the USSR) have signed the convention, and some consortia spokesman by no means dismiss the feasibility of operating under it as it now stands.

The potential significance of an international sea-bed authority cannot, however, be measured solely, or even mainly, in economic terms. Both in their content, and in the way in which they were arrived at, the "regime and machinery" for the international area of the sea-bed embodied in the Law of the Sea Convention constitute a novel step in the political organisation of this planet. For all their deficiencies, they offer scope for a world wide collaboration, not just in the regulation of mining, and the equitable redistribution of what it may yield, but, through the Enterprise, in the act of mining itself. Wisely, the Enterprise has not been invested with the sole right to mine, though the convention furnishes it with some advantages over its rivals, mining states, and private mining consortia. Should it, unhappily, squander those advantages, it would be a setback, but not a disaster. The convention does not put all its eggs, or even nodules, into one basket. Failure of the Enterprise would not impair (and might even enhance) the opportunities open to contractors to make a success of ocean mining. Its success, however, would enlarge the positive incentives for global co-operation beyond any that now exist and establish an organisation unlike any other, whose deliberations and decisions, because they dispose of real resources, would tend to lose the rhetoric which goes with the enunciation of principles, and acquire something of the practicality, and worldly pressures for compromise, that we find in domestic political processes. We have seen something of this in the collective workings of OPEC, though, unlike the Sea-bed Authority, that is an association of producers only, rather than a mix of interests, and it is not, as an institution, in business, on any large commercial scale. If a Sea-bed Authority mines the sea-bed, we shall all, through our states, if not otherwise, be shareholders; and that, politically, could give us a new conception of ourselves.

If the above analysis is right, the contribution a regime and machinery might make to future world order would itself be an overriding argument, for all who have interests that war could imperil (and who does not have?) for adherence to it. For the ocean mining states, who, for the most part, are the ones who are hesitating to commit themselves, the necessity of establishing a universally recognised title to mine what they undoubtedly view as a global commons is an added practical reinforcement. The burdens of mining under the regime of the convention may be considerable, but the hazards of mining outside it are surely intolerable.

Even if these arguments are disregarded, there remains a further point to merit reflection: the manner in which UNCLOS III has negotiated the Law of the Sea Convention and the assumptions underlying the process. From Caracas onwards, the professed aim of almost all delegates at UNCLOS III, and none more so volubly than those of Western States, was the negotiation of a "package deal". The package was: agree on a convention by which ocean mining states, who are also, by and large, the major maritime powers, are reassured about the effects of the coastal state revolution (whose necessity none would probably now dispute) on their distant water activities, - in other words, in which, as maritime states, they get, essentially, what they want, and they will agree to do, what, as mining states, only they can do, and make "the common heritage of mankind" a reality. Thus the regime and machinery for the sea-bed is now firmly implanted within the Law of the Sea Convention as a whole, and not detachable from it. States can obtain the benefit of the latter only be accepting the former.

Against this, it is now said, the rest of the convention has become customary international law. Everyone will look to the convention to determine the rights and duties of coastal states with respect to fishing or navigation within its jurisdiction, the rights of passage through straits, and so on. Therefore, a state does not need to become a party to the convention to enjoy these now "customary" rights. Even if this were true (and it ignores some rights, such as those of access to dispute settlement procedures, which are necessarily restricted to the convention's parties, except by agreement), to act on it would represent a massive, and damaging, act of bad faith on the part of those who called for a package deal in 1974. A regime and machinery for the international area was always part of the price, from the

Western point of view, which they offerred to pay for such a package. Not of course, any old regime, but one which was negotiated (as this was, up to 1980) and preserved the rights of access of ocean miners on reasonable conditions, as all including the West conceded that the draft convention (Rev 3) of 1980 did. If Western states proceed to mine, or encourage their companies to mine, the sea-bed in disregard of the convention, they will have gone back on the agreed raison d'être, at Caracas in 1974, of the attempt to arrive at a single comprehensive convention, and they will advertise, to the Group of 77, that the way to the creation of a more equitable world order does not lie through negotiations aiming at consensus. Of course developing countries, like everyone else at UNCLOS III, have pursued directly national concerns as well as, and no doubt more ardently than, conceptions of global justice; but the means matter. If the laborious process by which, in 1980, something very near a compromise among all those interests and conceptions was attained, is now blithely repudiated, the whole North-South dimension of international relations will be polarised; and the prospects for all subsequent negotiations thereby damaged. The attempt of UNCLOS III to give shape to the eloquent notion of the "common heritage of mankind", has set a test, as yet incomplete, whose significance is easy to observe but could, in the long run, prove momentous, of the prudence and statesmanship of contemporary governments and above all those of ocean mining states.

NOTES

1. I owe this term to Dr. Alexandra Post.

2. UNCLOS III, Provisional Verbatim Record, 190th Meeting, Montego Bay, December 8th 1982, (A/Conf. 62/PV.190) p.22.

3. Citizens for Ocean Law, Recent Events in the Law of the Sea, October 15th 1982.

Bibliography

Adams, F.G., 'The Impact of Cobalt Production from the Sea Bed', (UNCTAD, TD/B), (XIII)/MISC.3, 31st July 1973.

Alexander, L.M. (ed), 'The Law of the Sea: A New Geneva Conference', Proceedings of the Sixth Annual Conference of the Law of the Sea Institute, University of Rhode Island, Kingston, R.I., June 21st - 24th, 1971.

Amacher, R.C. and Sweeney, R.J. (eds), 'The Law of the Sea: U.S. Interests and Alternatives', American Enterprise Institute for Public Policy, Washington, D.C. 1976.

Amos, A.F. and Roels O.A., 'Environmental Aspects of Manganese Nodule Mining', Marine Policy, 1.2. April 1977.

Archer, A.A., 'Progress and Prospects of Marine Mining', Paper presented at the fifth Annual Offshore Technology Conference, Houston, Texas. April 29th - May 2nd 1973.

Archer, A.A., 'Resources and Potential Reserves of Nickel and Copper in Manganese Nodules', in Manganese Nodules: Dimensions and Perspectives, Proceedings of a United Nations Expert Group Meeting on Sea Bed Mineral Resources Library, Vol.II D.Reider Publishing Co., Dordrecht, 1979.

Auburn, F.M., 'Legal Aspects of Nodule Mining', in Glasby, G.P., (ed), Marine Manganese Deposits.

Barston, R.P. and Birnie. P.,(eds), 'The Maritime Dimension', Allen and Unwin, London 1980.

Bischoff, J.L. and Manheim, F.T., 'Economic Potential of the Red Sea heavy metal deposits', in Degens, E.T., and Ross, D.A., (eds) Hot Brines and Recent Heavy Metal Deposits in the Red Sea: A Geographical and Geophysical Account. Springer Verlag, 1969.

Boin. U. and Müller.E., 'Economic Aspects of Manganese Nodule Deep Ocean Mining', in Manganese Nodules, Metals from the Sea, Metallgesellschaft AG Review of Activities 18 1975.

Borgese, E.M. and Ginsburg, N. (eds), 'Ocean Yearbook 2', University of Chicago Press, 1979.

Bull, H., 'The Anarchical Society', Macmillan, London, 1977.

Buzan, B., 'Seabed Politics', Praeger, New York, 1976.

Buzan, B., 'United We Stand - informal negotiating groups at UNCLOS III', in Marine Policy, 4.(3), July 1980.

Buzan, B. and Middlemiss, D.W., 'The Exploitation of the Seabed' in Johnson, B. and Zacher, M.W., (eds), Canadian Foreign Policy and the Law of the Sea, University of British Columbia Press, Vancouver, 1977.

Christy, F.T., Clingan, T.A., Gamble, J.K., Knight, H.G. and Miles, E. (eds), 'Law of the Sea: Caracas and Beyond', Law of the Sea Institute Ninth Annual Conference, January 6th - 9th, 1975, Ballinger, Cambridge, Mass., 1975.

Citizens for Ocean Law, 'Recent events in the Law of the Sea', October 15th, 1982.

Commonwealth Secretariat, 'The North-South Dialogue: Making it Work', Commonwealth Secretariat, 1982.

Conahan, F.C., 'Financial Accountability of Proposed International Sea-bed Organisation', Case study presented to the U.S. Department of State Senior Seminar in Foreign Policy, Foreign Service Institute, 1973.

Crawford, Sir J. and Saburo Okita, D.R., 'Australia, Japan and Western Pacific Economic Relations', Australian Government Publishing Service, 1976.

Cronan, D.S., 'Manganese Nodules and Other Ferro-manganese Deposits From the Atlantic Ocean', Journal of Geophysical Research, 80 (27), September 20th, 1975, pp.3831-3837.

Cronan, D.S., 'Riches of the Ocean Floor' in SPECTRUM, British Science News, (1976), No.137/1, pp.10-12.

Cronan, D.S. and Tooms, J.S., 'The geochemistry of manganese nodules and associated pelagic deposits from the Pacific and Indian Oceans', Deep-Sea Research, 16, 1969, pp.335-359.

Doumani, G.A., 'Ocean Wealth: Policy and Potential', Hayden Book Co., Rochelle Park. N.J., 1973.

Dubs, M., 'An Industrialist's Reaction to The Law of the Sea Conference', remarks before the Southern Center for International Studies, May 24th, 1978.

Eckert, R.D., 'The Enclosure of Ocean Resources', Stanford University Press, 1978.

Ely, N., 'A Deep Ocean Mining Claim', paper presented to the American Mining Convention, September 28th - October 1st 1975.

Eustis, R.D., 'Procedures and Techniques of Multinational Negotiation', Virginia Journal of International Law, 1977.

Evensen, J., 'Informal consultation in Geneva, 28th February - 11th March, 1977, on matters relating to the First Committee of the Law of the Sea Conference, in particular the system of exploitation', Report addressed to the President of the Conference and Chairman of the First Committee, 25th April, 1977.

Flipse, J., 'Deep Ocean Mining Technology and its Impact on the Law of the Sea', in Christy, F. and others (eds), Law of the Sea: Caracas and Beyond. Ballinger, Cambridge, Mass., 1975.

Forsyth, M.G., Keens-Soper, H.M.A. and Savigear, P. (eds), 'The Theory of International Relations', Allen and Unwin, London, 1970.

Gamble, J.K. (ed), 'Law of the Sea: Neglected Issues', Proceedings of the Law of the Sea Institute Twelth Annual Conference, The Hague, October 23rd - 26th, 1978 Law of the Sea Institute, Hawaii, 1979.

Gamble, J.K. and Pontecorvo, G.(eds), 'Law of the Sea: The Emerging Regime of the Oceans', Proceedings, Law of the Sea Institute Eighth Annual Conference, June 18th - 21st, 1973, Kingston, R.I., Ballinger, Cambridge, Mass., 1974.

Garvey, G. and Garvey, L.A.(eds), 'International Resource Flows', Heath and Co., Lexington, Mass., 1977.

Gaskell, T.F. and Simpson, S.J.R., 'Oil 2 billion B.C. to A.D. 2000', in Borgese, E.M. and Ginsburg, N.(eds), Ocean Yearbook 2, University of Chicago Press, 1979.

Gauthier, M. and Marvaldi, J., 'The Two-ship CLB System for Mining Polymetallic Nodules', paper given to Oceanology International 75 Conference, Brighton 1975.

Glasby, G.P., (ed) 'Marine Manganese Deposits', Elsevier, Amsterdam, 1977.

Grotius, H., 'Prolegomena to the three books on the Law of War and Peace.'

Harris, S., 'The Commodities Problem and the International Economic Order', in Oppenheimer, P.(ed), Issues in International Economics, Oriel Press, 1980.

Hasegawa, K., 'History and Outlook of the Development of Production Technology of Manganese Nodules', in The Deep Seabed and its Mineral Resources: Proceedings of the third International Ocean Symposium, Tokyo, 1978.

Heezen, B.C., Tharp, M. and Ewing, M., 'The Floors of the Oceans', The Geological Society of America, Special Paper 65, April 11th, 1959.

Hjertonsson, K., 'The New Law of the Sea', Sijthoff, Leiden, 1973.

Hobbes, T., 'Leviathan'.

Hodgson, R.D., 'National Maritime Limits: the Economic Zone and the Seabed', in Christy, F.T. and others (eds), Law of the Sea: Caracas and Beyond, Ballinger, Cambridge, Mass., 1975.

Hoffmann, S., 'The State of War', Pall Mall, London, 1965.

Hollick, A.L., 'Bureaucrats at Sea', in Hollick, A.L. and Osgood, R.E., New Era of Ocean Politics, Johns Hopkins Press, Baltimore, 1974.

Hollick, A.L., 'US Foreign Policy and the Law of the Sea', Princeton U.P., 1981.

Hollick, A.L., 'The Truman Proclamations', Virginia Journal of International Law, Fall, 1976.

Hollick, A. L., 'U. S. Foreign Policy and the Law of the Sea', Princeton U. P., 1981.

Johnson, B. and Zacher, M.W. (eds), 'Canadian Foreign Policy and the Law of the Sea', University of British Columbia Press, Vancouver, 1977.

Johnson, D.B. and Logue, D.E., 'U.S. Economic Interests in Law of the Sea Issues', in Amacher, R.C. and Sweeney, R.J. (eds), The Law of the Sea: U.S. Interests and Alternatives, American Enterprise Institute for Public Policy, Washington, D.C., 1976.

Johnson, D.H.N., The North Sea Continental Shelf Cases, International Relations November 1969, pp.522-540.

Johnston, D.M.(ed), 'Marine Policy and the Coastal Community', Croom Helm, 1976.

Keohane, R.O. and Nye, J.S., 'Power and Interdependence', Little, Brown and Co., Boston, 1977.

Kirthisingha, P.N., 'The Enterprise - an expendable triumph', Marine Policy, 5.3, July 1981.

Kissinger, H.A., 'The Law of the Sea: A Test of International Cooperation', address to the Foreign Policy, U.S. Council of the International Chamber of Commerce, and U.N.A. of U.S.A., April 8th, 1976.

Kissinger, H.A., 'The Laws of the Sea and Space - Vital Issues in an International World', address to American Bar Association, Montreal, August 11th, 1975.

Krasner, S., 'The Quest for Stability, Structuring the International Commodities Markets', in Garvey, G. and Garvey, L.A. (eds), International Resource Flows, Heath and Co, Lexington, Mass., 1977.

Kruger, J. and Schwarz, K.H., 'Processing of Manganese Nodules', in Manganese Nodules Metals from the Sea, Metallgesellschaft A.G., Review of Activities 18, 1975.

Kubalkova, Y. and Cruickshank, A.A., 'A Double Omission', British Journal of International Studies, 3.(3), 1977.

Leipziger, D.M. and Mudge, J.L., 'Seabed Mineral Resources and the Economic Interests of Developing Countries', Ballinger, Cambridge, Mass., 1976.

Luard, E., 'The Control of the Sea-Bed', Heinemann, London, 1974.

Machiavelli, N., 'The Prince', Everyman's Library, Dent and Sons, London, 1958.

McKelvey, V.E. and Wright, N.A., 'A Preliminary Analysis of the World Distribution of Subsea Metal-rich Manganese Nodules', [paper presented to the Neptune Group seminar at UNCLOS III, Ninth Session, Geneva, past, August, 1980, pp.8-5].

Malenbaum, W., 'World Demand for Raw Materials in 1985 and 2000' (Philadelphia, Wharton School of Finance and Commerce, University of Pennsylvania, 1977).

Malone, J.L.(U.S. Ambassador to UNCLOS III), 'Statement to Plenary, April 1st, 1982,' Press Release USUN 14 (82), April 1st, 1982, U.S. Mission to the United Nations.

Menard, H.W. and Frazer, J.N., 'Manganese Nodules on the Sea Floor: Inverse Correlation Between Grade and Abundance', Science, 199 (1978), pp.969-970.

Mero, J.L., 'The Mineral Resources of the Sea', Elsevier, Amsterdam, 1964.

Mero, J.L., 'The Great Nodule Controversy', in Christy, F.T. and others (eds), Law of the Sea: Caracas and Beyond, Ballinger, Cambridge, Mass, 1975.

Metallgesellschaft A.G., Review of Activities, 18 (1975), 'Manganese Nodules, Metals from the Sea'.

Miles, E., 'An Interpretation of the Caracas Proceedings', in Christy, F.T. and others (eds), Law of the Sea: Caracas and Beyond, Ballinger, Cambridge, Mass., 1975.

Miles, E., 'The Dynamics of Global Ocean Politics', in Johnston, D.M.(ed), Marine Policy and the Coastal Community, Croom Helm, London, 1976.

Miles, E. and Gamble, J.K. (eds), 'Law of the Sea: Conference Outcomes and Problems of Implementation', Proceedings, Law of the Sea Institute Tenth Annual Conference, June 22nd - 25th, 1976, Kingston, R.I., Ballinger, Cambridge, Mass., 1977.

Mitchell, B., 'A Future Regime for the Seabed', M.A. Thesis, University of Sussex, 1974.

Oda, S., 'The International Law of the Ocean Development I', Sijthoff, Leiden, 1972.

Oda, S., 'The International Law of the Ocean Development II', Sijthoff, Leiden, 1975.

Oda, S., 'The Law of the Sea in our Time II, The United Nations Seabed Committee, 1968-73', Sijthoff, Leiden, 1977.

Oda, S., 'Proposals for Revising the Convention on the Continental Shelf', in Columbia Journal of Law, 7, 1968.

Odell, P., 'Offshore Resources: Oil and Gas', in Barston, R.P. and Birnie, P. (eds), The Maritime Dimension, Allen and Unwin, London, 1980.

Ogley, R.C., 'Caracas and the Common Heritage', International Relations IV(6), November 1974, pp.604-628.

Ogley, R.C., 'Decision-making in the United Nations - the case of the Representation of China', International Relations, April, 1964.

Ogley, R.C., 'The Law of the Sea Draft Convention and the New International Economic Order', Marine Policy 5(3), July 1981, pp.240-251.

Ogley, R.C., 'The United Nations and East-West Relations, 1945-71' I.S.I.O., Monograph, University of Sussex, Brighton, 1972.

Ogley, R.C., 'Towards a General Theory of International organisation', International Relations III(8), November, 1969.

Oppenheimer, P.(ed), 'Issues in International Economics', Oriel Press, 1980.

Page, W. and Rush, H., Long-term Forecasts for Metals: the Track Record 1910-1960's, Occasional Paper No. 6, Science Policy Research Unit, University of Sussex, 1978.

Pardo, A., 'The Common Heritage: Selected Papers on Oceans and World Order 1967-1974', International Ocean Institute, Occasional Papers 3, Malta University Press, 1975.

Pardo. A. and Borgese, E.M., 'The New International Economic Order and the Law of the Sea', International Ocean Institute, Occasional Papers 4, Malta University Press, 1976.

Post, A., 'Deepsea Mining and the Law of the Sea', Martini, Nijhoff, The Hague, 1983.

Prain, Sir R., 'Copper: the Anatomy of an Industry', Mining Journal Books, London 1975.

Prescott, J.R.V., 'The Political Geography of the Oceans', David and Charles, Newton Abbot, 1975.

Quaker Office at the United Nations, 'Five Panel Discussions on UNCLOS III: Summary of Remarks of Speakers', Quaker United Nations Office, Geneva, 1978.

Ratiner, L., 'Reconstructed Extemporaneous Statement before informal meeting of First Committee, August 9th, 1974'.

Ratiner, L., 'The Law of the Sea: A Crossroads for American Foreign Policy', Foreign Affairs, Summer 1982.

Reidel, O., 'Mining and Transport of Manganese Nodules - Technological Problems', in Metals from the Sea, Metallgesellschaft AG Review, 18, 1975.

Richardson, E.(U.S. Ambassador to UNCLOS III), 'Comments on Financing', March 10th, 1977.

Richardson, E.(U.S. Ambassador to UNCLOS III), 'Statement', May 22nd, 1978.

Richardson, E.(U.S. Ambassador to UNCLOS III), 'Statement in plenary', August 26th, 1980.

Ross. D.A., 'Resources of the Deep Sea Other than Manganese Nodules', in Gamble, J.K.(ed), Law of the Sea: Neglected Issues, Law of the Sea Institute, Hawaii, 1978.

Russell, R.B., 'The United Nations and United States Security Policy', Brookings Institution, Washington, D.C., 1968.

Siapno, W.D., 'Undersea Assessment' summarized in _Metal Bulletin Monthly_, November 1975, pp.35-36.

Smith, B., 'Long-term Contracts in the Resource Goods Trade', in Crawford Sir J. and Saburo Okita, D.R., _Australia, Japan and Western Pacific Economic Relations_, Australian Government Publishing Service, 1976.

Smith, D.N. and Wells, L.T., '_Negotiating Third World Mineral Agreements_', Ballinger, Cambridge, Mass., 1975.

Spero, J., '_The Politics of International Economic Relations_', Allen and Unwin, London 1977.

Sreenivasa Rao, P., '_The Public Order of Ocean Resources_', MIT Press, Cambridge, Mass., 1975.

Stevenson, J.R.(U.S. Ambassador to UNCLOS III), 'Statement to Plenary', July 11th, 1974, Department of State Press Release No.4.

Stevenson, J.R.(U.S. Ambassador to UNCLOS III, Caracas), 'Statement to the First Committee', July 17th, 1974, Department of State Press Release No.5.

United Nations, '_Economic Implications of Seabed Mineral Development in the International Area: Report of the Secretary General_', UN DOC. A/Conf. 62/25, May 22nd, 1974.

United Nations Department of Economic and Social Affairs, '_Sea-bed Mineral Resource Development: Recent Activities of the International Consortia_', UN DOC. ST/ESA/107, 1980.

United Nations General Assembly, '_Secretary General's Report on the Economic Significance in Terms of Sea-bed Mineral Resources of the Various Limits Proposed for National Jurisdiction_', UN DOC A/AC 138/87 June 4th, 1973.

United Nations Secretary-General, '_Additional Notes on the Possible Economic Implications of Mineral Production from the International Sea-Bed Area_', UN DOC. A/AC 138/73, May 1972.

United Nations Secretary-General, '_Sea-bed Mineral Resources: Recent Developments: Progress Report_' UN DOC A/AC 139/139 July 1973.

United Nations 'Third United Nations Conference on the Law of the Sea', _Official Records_.

U.S. Bureau of Mines, '_Mineral Facts and Problems_', 1975 edition, Bulletin 667, U.S.Government Printing Office, 1976.

U.S. Congress, House of Representatives Committee on

International Relations, Subcommittee on International
organisation: 'Deep Seabed Minerals: Resources Diplomacy
and Strategic Interest', U.S. Government Printing Office,
March 1978.

U.S. Delegation to UNCLOS III, 'Approaches to Major Problems
in Part XI of the Draft Convention on the Law of the Sea',
February 24th, 1982.

U.S. Delegation to UNCLOS III, 'Report Sixth Session', May
28th - July 15th, 1977.

U.S. Delegation to UNCLOS III, 'Report Eighth Session',
Geneva, March 19th - April 27th, 1979.

U.S. Delegation to UNCLOS III, 'Report Ninth Session', March
3rd - April 4th, 1980.

Willetts, P., 'The Non-Aligned Movement', Frances Pinter,
London, 1978.

World Energy Resources 1985-2020. World Energy Conference,
IPC, Guildford 1978.

Wright, R.L., 'Ocean Mining: An Economic Evaluation',
Professional Staff Study of the Ocean Mining
Administration, U.S.Department of the Interior, May
1976.

Yearbook of the International Law Commission, 1950, Vol 1,
'Summary Records of the Second Session', June 5th - July
29th, 1950.

Index

Abyssal plain 4-5
Adjacency 105, 129
AFERNOD 44
Aguilar, Andres 69, 71
Airlift hydraulic dredgers . . 14
Albania 94
Algeria 95-96
Amerasinghe, Hamilton Shirley 65, 71, 77, 79-81, 88-89
Amoco Ocean Minerals Co. . . . 45
AMR 45
Anti-monopoly clause 155, 172, 188, 231
Archer, Alan 10
Archipelagic Group 84
Archipelagoes 101, 124
    Economic Zone 125
    Sea-Bed Committee 125
Arena principle 70-71, 90
Argentina 95
Asian-African Legal Consultative Committee 113-114, 129
Assurance of Access 152, 154, 156, 163-164, 166,
                                169, 171, 237
Atlantic Ocean 7
Australia 20-22, 48, 96

Bahrain 95
Baselines 99
    straight 101
Basic Conditions of Exploitation 143-144, 146, 150, 153,
                                174, 184, 210, 229
Belgium 1, 46, 79
Billiton 45
Binding commercial arbitration 221
Brazil 24
Brazil Clause 162-163
British Petroleum 44
Bureau de Recherches Geologiques et Minieres - BRGM 44
Buzan, B. 44-45, 84

Canada 6, 20-22, 46, 48, 51, 82,
                                95-96
Carter Administration 74, 240
Challenger expedition of 1873-6 12
Chantiers de France - Dunkerque 44
Chile 20, 96
China 35-36, 70
Clarion-Clipperton zone . . . 8
CNEXO 44
Coastal states 46-48
Cobalt 7-9, 11, 19, 22-23, 30
Commissariat a l'energie . . . 44
Commodity agreements 185-187, 189, 191-192

Common Heritage Fund 130
Common heritage of mankind . . 68
Commonwealth 51
Commonwealth Group of Experts 87
Commonwealth Secretariat . . . 83
Complex interdependence . . . 40
Consolidated Gold Fields . . . 44
Continental margin 4, 100, 103, 107, 116
Continental rise 4, 100, 129
Continental shelf 108, 129, 237
Continental Shelf Convention (1958) 5, 104, 106, 237
Continental slope 4-5, 108, 129, 237
Continuous line bucket system 13, 18, 25, 44-45
Copper 7-9, 11, 18-20, 26-27, 30
Cuba 21
Cybernetic theory 40
Cyprus 20

De Soto, Alvaro 83
Deep Ocean Minerals Association 44
Deep Ocean Mining Company (DOMCO) 45
Deepsea Ventures 12, 14, 19, 23-25, 29, 44-45,
                               52, 135
Dispute-settlement mechanisms 149, 171, 215
Dominican Republic 21
Dredge efficiency 15
Dual system 147
Dubs, Marne 18, 29, 46

East Pacific Rise 27
Echo sounders 13
Economic Zone 96, 100, 103, 108, 114-116,
                               237
Economic Zone of 200 miles . . 68
Ely, Northcutt 12, 29, 46, 52
Engo, Paul 64, 69-70, 72-77, 80, 85, 96
ENI Group 45
Enterprise, the 142, 145, 147-148, 150, 159,
                               161-166, 171-174, 190-191,
                               193, 197-198, 211-213,
                               226-229, 231-234, 237, 241,
                               243, 245
   first site guarantees for . 164-165
Equitable geographical distribution 199, 205, 211-212
European Economic Community (EEC) 51, 78, 82
Evensen group (1972-1975) . . 71, 84-85
Evensen, Jens 69, 72-73, 80, 85
Exploitability criterion . . . 105-106

Federal Republic of Germany . 1, 46, 79
Ferro-manganese 23
Financial arrangements 157
Financial obligations of contractors 81, 83, 156-161, 172
Flags of convenience 228
Flipse, John 29, 45-46
France 46, 79

Freefall grab samplers 13
Friends of the Conference (1982)   85, 92, 210, 222, 226,
                          228, 243
Friends World Committee for Consultation   51
Front-end loading 159
Functionalism 38

Gabon 23, 45
Galapagos rift 27
GATT Rules 190
Gentlemen's agreement 62
Geographically-disadvantaged states   48, 68, 81, 83
Ghana 24
Ghana-Guinea Union 37
Global Marine 44-45
Grab samplers 13
Group of 11 210
Group of 77 . . . . 47-49, 54-55, 57-58, 72-74,
                          76-82, 84, 87, 89, 91-92, 97,
                          110, 113-114, 143, 146-148,
                          151, 153, 159, 165, 173-175,
                          184, 186, 189-190, 192, 196,
                          198, 200-207, 209-211, 219,
                          225-233, 238-241, 243, 247
Guatemala 95
Gulf of Mexico 5

Historic bays 101
Hobbes 31
Hughes, Howard 45
Hydrocarbons 5
Hydrometallurgy 17
  techniques,use of 26

Iceland 96
INCO 44-45
India 23, 46, 70, 96
Indian Ocean 7
Indonesia 21, 96
Informal Composite Negotiating Text (ICNT)   72, 80, 204, 218
Interim period 187-188, 237-238
International Court of Justice (ICJ)   102, 107, 216-217, 242
International Law Commission . 105
International sea-bed boundary   63
Islands 125-127
Italy 1, 46, 79

Jackling, Sir Roger 95
Japan 46, 79, 82
Japanese Manganese Nodule Development Company (JAMCO)   45
Joint ventures 150-151, 154, 169, 172-173,
                          243
Juan de Fuca Ridge 27
Judicial review 196-197, 222

Kennecott Group 18, 29, 44, 46

Kissinger, Henry 74
Koh, Tommy 69, 77, 81, 159-160

Land-based producers 47-48, 50, 82, 180, 182-186,
                                190-191, 200, 206, 209, 241
  actual and potential 46
Land-Locked and Geographically-Disadvantaged States (LLGDS)
                                84
Land-locked states 47-48, 50, 68, 81, 83
Law of the Sea Tribunal . . . 218-221, 223, 235
  Sea-bed Disputes Chamber . . 211, 218-219, 221-222
Least-developed countries . . 50
Lima Declaration 112, 114
Lockheed Corporation 44-45
London Dumping Convention . . 26
LOS (Law of the Sea) Convention 246
  adopted, April 30 1982 . . . 1
  Annex III 6
  Annex IV 6
  formalisation of text (1982) 78
  Part XI 6
  participation 78
  signed, Montego Bay, December 1982 1
Luard, Evan 36, 82

Machiavelli 31
Malta 32, 56, 60
Manganese 7-9, 11, 19, 23-24, 30
  manganese markets 23
Manganese nodules 6
  metal content and depth . . 8
Marconaflo 16
Marxism 34
Mauritius 96
Mero, John 4, 7, 9, 11-12, 18-19, 29-30
Metal-exporting states (see land-based producers) 46
Metal-importing states 46
Metallgesellschaft 15, 19, 45
Mexico 96
Mitsubishi Corporation 44
Molybdenum 7, 19, 24, 27, 30
Mongolia 95
Montevideo Declaration 112, 129
Morocco 24

Nandan, Satya 76, 81
National legislation 43, 225, 241-242
NEPTUNE 51
Netherlands 46, 70
New Caledonia 21-22
New Zealand 96
Newport News Shipbuilding and Drydock Company 45
Nickel 7-9, 11, 18-19, 21-22, 30
Nickel market 21
NIEO (New International Economic Order) 49, 86-87, 97, 183
  Declaration and Action Programme 97

Njenga, Frank 69, 76, 187
Nodule mining
  attributed net proceeds of . 157-160, 178
  environmental aspects of . . 25-27, 180
  profitability of 19
  refining process 29
  size of sites 11, 15
Nodule production
  impact on copper market . . 20
Nodules, promising ocean areas 10
Nodules, reserves of minerals in 11
Non-Governmental Organisations 51, 70
Non-nodule resources 150
Noranda Mines 44
North Sea Continental Shelf Cases of 1969 102
North Vietnam 62
Norway 74

Ocean Education Project . . . 51
Ocean Management Incorporated 45
Ocean Minerals Company 45
Ocean Minerals Inc. 45
Ocean Mining Associates . . . 45
Ocean Resources 44
Ocean space 143
Oil industry 52
Oil price rise 1973 85
Organization of African Unity (OAU) 116

Pacific Ocean 7
Panama Declaration 109
Paraguay 95
Parallel system 134, 138, 145-146, 148,
                                150-151, 156, 162-164, 167,
                                170, 184-185, 234, 237
Pardo, Arvid 4, 32, 34, 55, 110, 138
Patrimonial Sea 114-115
Peru 20, 73, 95
Philippines 20
Pinto, Christopher 64, 69, 72, 81
Pioneer Area 230-232, 238
Pioneer investors 227, 231, 238
Pioneer Operator 227
Pioneer status 227, 230, 232
PLO 78
Pneumatic sound generators . . 13
Political functionalism . . . 39
Polymetallic sulphides 26-27
Potential producers 185
Preparatory Commission 78, 153, 170, 190, 206, 224,
                                227-231, 233-234, 236-238,
                                241-243
Preparatory Investment Protection (PIP) 78, 225, 234-235
Production control 180-183, 185, 187-188, 192
Production limit 184-186, 189, 192, 228-229,
                                231

Prospective sea-bed miners . . 43
Pyrometallurgy 17
   techniques,use of 26

Quasi-compromise of 1980 . . . 134, 149-151, 156-170

Ratiner, Leigh 46, 74, 91, 154
Reagan Administration 77, 81, 91, 170, 207
Reagan Review 77, 81, 134, 161, 190, 192,
                               211, 213, 221, 225, 234
Reagan, Ronald 38
Red Sea metalliferous muds . . 27
Revenue-sharing 207
Revenue-sharing by coastal states 116-118, 130
Reversibility 40, 89
Review conference 166-175, 179, 191, 206, 210,
                               214, 238, 244
Revised Single Negotiating Text (RSNT) 83, 202
Richardson Review (1977-8) . . 75
Richardson, Elliot 46, 74, 96, 154
Rio Tinto Zinc 44
Romania 95
Rousseau 32
Royal Boskalis 45
Royal Dutch/Shell Group . . . 45

Salt-domes 5
Salzgitter 45
Sanim 45
Santiago Declaration 109
Sea-Bed Authority
   Assembly 152, 165, 174, 197, 199-202,
                               204, 206-209, 212-213,
                               218-219, 221, 234, 243-244
   Council 150, 152, 154, 165, 170-171,
                               173-174, 191, 197-213,
                               217-219, 221, 227, 233-234,
                               243-244
   Enterprise, the 142, 145, 147-148, 150, 159,
                               161-166, 171-174, 190-191,
                               193, 197-198, 211-213,
                               226-229, 231-234, 237, 241,
                               243, 245
   Legal and Technical Commission 155, 171, 173, 206, 208,
                               210-211, 243
   Secretary-General 202
   Tribunal 201-202, 216-219
Security of tenure 144, 151, 184
SEDCO (USA) 45
Self-positioning drill ships . 5
Sestonophages 25
Shelf-locked states 47-48
Shultz, George 51
Silicon 7
Single Negotiating Text (SNT) 149-150, 184
Site-banking 146-147

Societe le Nickel 44
South Africa 20, 24, 45
South Korea 105, 112, 129
South Vietnam, Provisional Revolutionary Government of 62
Soviet Union 58, 65
Spade corers 13
Sri Lanka 64, 80
Standard Oil of Indiana . . . 45
Standard Oil of Ohio 44
Sub-sea completion systems . . 5, 127
Sumitomo Group 44
Summa Corporation 44-45
Sun Co. 45
Superjacent waters 107
Sweep efficiency 14

Tenneco 45
Territorial sea 108-109, 111, 123, 127
  conference proposed on . . . 58
Territorialist Group 81
Thatcher, Margaret 38
Title to mine 242-243, 246
Title to sites, need for . . . 135-136
Transfer of technology 148, 154, 156, 162-163,
                                171-175, 227, 230
Transverse thruster plants . . 16
Truman Proclamation 104-105, 107, 109, 127
Turbidity 25

UK 1, 46, 79, 82
UN (United Nations) 55, 244
  Charter 195
  diplomacy 57
  Economic and Social Council 93
  forces
    in Cyprus 39
    in the Congo (ONUC) . . . 39
  General Assembly 32, 55, 59-61, 115, 138,
                                195-197, 207, 236, 242
    Ad Hoc Sea-Bed Committee . 2, 59, 79, 94, 182
    Declaration of Principles (1970) 1-2, 61, 68, 112,
                                138-140, 146, 182, 198, 216
    First Committee 2, 61, 115, 182
    Moratorium Resolution (1969) 61, 138-139
    Sea-Bed Committee 2, 49, 54, 59-60, 62, 64, 79,
                                81, 84, 94, 112, 114-115,
                                128, 137-138, 142, 196-198
    Seventh Special Session (1975) 86
    Sixth Special Session (1974) 86, 183
  International Law Commission 104-105
  Secretariat 56, 93
  Secretary-General 60-61, 102, 107, 115, 160,
                                182-183, 190, 195
  Security Council 195
UNCITRAL Arbitration Rules . . 221
UNCLOS I (1958) 63, 66, 104, 109, 199

UNCLOS II (1960) 66, 111, 199
UNCLOS III (1973-82) 32, 49, 54, 65-66, 103, 109,
                                129, 137, 168, 175, 183, 187,
                                190, 194, 198-199, 217,
                                235-236, 238-239, 241, 244,
                                246-247
  agenda of 55
  collegiate principle 72
  Commission on the Limits of the Continental Shelf 119
  continental margin
    Economic Zone and 122
    in Negotiating Group 6 . . 118, 120
    in Second Committee . . . 119-120
    IOC's comments on 120
    Irish formulae for 119-122
    legal definition of . . . 118-122
    oceanic ridges and 122
  cooling off provisions . . . 66
  Draft Conventions (1980 _1981) 91, 97, 205
  duration of 54
  First Committee 51, 65, 67, 69-70, 72-73, 75,
                                80-84, 98, 139-140, 142-143,
                                145-146, 151, 162, 166, 174,
                                179, 183, 186-187, 200, 203,
                                205, 207, 215-220
    Ad Hoc Negotiating Group (1976) 70
    Chairman's Working Group (1977) 69, 73, 80, 166,
                                186-187, 203, 218
    Group of 21 . . . . 70-71, 76, 90, 205
    Informal Workshop 70
    Working Group 201
  First Session (New York 1973) 62, 65
  Group of Legal Experts . . . 96
  informal meetings of the whole 68-70, 76, 79
  Informal Plenary ('fourth committee') 72, 215, 217-218
  inter-sessional talks (1977) 72, 80
  mandate of 60
  Negotiating Group 1 . . . . 154, 169, 187-188
    Sub-group of Technical Experts (Archer Group) 21
  Negotiating Group 2 . . . . 77, 81, 159, 161
  Negotiating Group 3 . . . . 77, 204
  negotiating groups (1978 on) 69
  negotiating texts 71
  package deal 63, 66
  Rules of Procedure 65, 79
  Second Committee 69, 71, 83-84, 96, 98, 123
    "Main Trends" 71
  Second Session (Caracas 1974) 22-23, 51, 54, 62, 65, 67,
                                69, 71, 79, 83, 91, 96, 116,
                                143
  Secretariat 69
  Third Committee 145
  Third Session (Geneva 1975) 69
UNCTAD 49, 51, 82, 86
UNCTAD 'Common Fund' 49, 87, 244
UNCTAD IV, Nairobi 87

UNCTAD Secretary-General . . . 87
UNEP 51
Union Miniere of Belgium . . . 45
Unitary system 148
United Arab Republic 37
US Department of Interior . . 52
US Steel 45
USA 1-3, 6, 20, 24, 32, 46,
                               49-50, 58, 73-74, 76, 78-80,
                               82, 88-89, 91-92, 148,
                               151-152, 175, 181, 183-184,
                               187, 190-191, 195-196,
                               201-202, 208-209, 212, 226,
                               236, 240, 242
   "Approaches" paper (1982) . 170-172, 190, 207, 210, 221
   Congress 51
   Defense Department 50, 91
   Draft Convention of 1970 . . 50, 107-108, 198, 217, 219
      International Trusteeship Area   108, 141
      redistributive potential . 140
      role of Tribunal in . . . 216
   fishing industry 50
   Green Book (1982) 172, 174, 179, 190, 208-209,
                               213, 221
   hard minerals industry . . . 51
   marine science community . . 50
   military 50
   petroleum industry 50
   sea-bed miners 50
   State Department 50
   Treasury 51
USSR 21, 24, 35-36, 38-39, 46, 77

Vanadium 8-9, 19, 24, 27, 30
Venezuela 80
Vienna formula 62

Weighted voting 49
Wuensche, Harry 76

Yankov, Alexander 145

Zaire 20, 22, 24
Zambia 20, 22
Zinc 9, 19, 24, 27